# Pseudo Limits, Biadjoints, and Pseudo Algebras: Categorical Foundations of Conformal Field Theory

# MEMOIRS
of the
American Mathematical Society

Number 860



# Pseudo Limits, Biadjoints, and Pseudo Algebras: Categorical Foundations of Conformal Field Theory


Thomas M. Fiore



July 2006 • Volume 182 • Number 860 (end of volume) • ISSN 0065-9266

**American Mathematical Society**
Providence, Rhode Island

2000 *Mathematics Subject Classification.*
Primary 18C10, 18C20; Secondary 81T40, 18A30.

**Library of Congress Cataloging-in-Publication Data**

Fiore, Thomas M., 1977–
Pseudo limits, biadjoints, and pseudo algebras : categorical foundations of conformal field theory / Thomas M. Fiore.
p. cm. — (Memoirs of the American Mathematical Society, ISSN 0065-9266 ; no. 860)
"Volume 182, number 860 (end of volume)."
Includes bibliographical references.
ISBN 0-8218-3914-4 (alk. paper)
1. Conformal invariants. 2. Quantum field theory. I. Title. II. Series.

QA3.A57 no. 860
[QC174.52.c66]
510 s—dc22
[530.14′3] 2006042824

## Memoirs of the American Mathematical Society

This journal is devoted entirely to research in pure and applied mathematics.

**Subscription information.** The 2006 subscription begins with volume 179 and consists of six mailings, each containing one or more numbers. Subscription prices for 2006 are US$624 list, US$499 institutional member. A late charge of 10% of the subscription price will be imposed on orders received from nonmembers after January 1 of the subscription year. Subscribers outside the United States and India must pay a postage surcharge of US$31; subscribers in India must pay a postage surcharge of US$43. Expedited delivery to destinations in North America US$35; elsewhere US$130. Each number may be ordered separately; *please specify number* when ordering an individual number. For prices and titles of recently released numbers, see the New Publications sections of the *Notices of the American Mathematical Society.*

**Back number information.** For back issues see the *AMS Catalog of Publications.*

Subscriptions and orders should be addressed to the American Mathematical Society, P. O. Box 845904, Boston, MA 02284-5904, USA. *All orders must be accompanied by payment.* Other correspondence should be addressed to 201 Charles Street, Providence, RI 02904-2294, USA.



*Memoirs of the American Mathematical Society* is published bimonthly (each volume consisting usually of more than one number) by the American Mathematical Society at 201 Charles Street, Providence, RI 02904-2294, USA. Periodicals postage paid at Providence, RI. Postmaster: Send address changes to Memoirs, American Mathematical Society, 201 Charles Street, Providence, RI 02904-2294, USA.



This publication is indexed in *Science Citation Index®*, *SciSearch®*, *Research Alert®*, *CompuMath Citation Index®*, *Current Contents®/Physical, Chemical & Earth Sciences.*
Printed in the United States of America.

∞ The paper used in this book is acid-free and falls within the guidelines established to ensure permanence and durability.
Visit the AMS home page at http://www.ams.org/

10 9 8 7 6 5 4 3 2 1 11 10 09 08 07 06

In memory of my Mother

# Contents

# Abstract

In this paper we develop the categorical foundations needed for working out completely the rigorous approach to the definition of conformal field theory outlined by Graeme Segal. We discuss pseudo algebras over theories and 2-theories, their pseudo morphisms, bilimits, bicolimits, biadjoints, stacks, and related concepts.

These 2-categorical concepts are used to describe the algebraic structure on the class of rigged surfaces. A *rigged surface* is a real, compact, not necessarily connected, two dimensional manifold with complex structure and analytically parametrized boundary components. This class admits algebraic operations of *disjoint union* and *gluing* as well as a *unit.* These operations satisfy axioms such as unitality and distributivity up to coherence isomorphisms which satisfy coherence diagrams. These operations, coherences, and their diagrams are neatly encoded as a *pseudo algebra over the 2-theory of commutative monoids with cancellation.* A *conformal field theory* is a morphism of stacks of such structures.

This paper begins with a review of 2-categorical concepts, Lawvere theories, and algebras over Lawvere theories. We prove that the 2-category of small pseudo algebras over a theory admits weighted pseudo limits and weighted bicolimits. This 2-category is biequivalent to the 2-category of algebras over a 2-monad with pseudo morphisms. We prove that a pseudo functor admits a left biadjoint if and only if it admits certain biuniversal arrows. An application of this theorem implies that the forgetful 2-functor for pseudo algebras admits a left biadjoint. We introduce stacks for Grothendieck topologies and prove that the traditional definition of stacks in terms of descent data is equivalent to our definition via bilimits. The paper ends with a proof that the 2-category of pseudo algebras over a 2-theory admits weighted pseudo limits. This result is relevant to the definition of conformal field theory because bilimits are necessary to speak of stacks.

---

Received by the editor April 2nd, 2004.

2000 *Mathematics Subject Classification.* Primary 18C10, 18C20; Secondary 81T40, 18A30.

*Key words and phrases.* 2-categories, pseudo limits, pseudo algebras, lax algebras, biadjoints, Lawvere theories, 2-theories, stacks, rigged surfaces, conformal field theory.

# Acknowledgements

It is with great pleasure that I acknowledge the many people who have aided me in the creation of this book. I am deeply grateful to Igor Kriz for his careful guidance and encouragement. My gratitude extends to Po Hu, F. W. Lawvere, Ross Street, Steve Lack, John Baez, Tibor Beke, Bob Bruner, James McClure, Jeff Smith, Art Stone, Martin Hyland, John Power, Michael Johnson, Mark Weber, Craig Westerland, and Bart Kastermans for helpful comments. Barbara Beeton enhanced the format of this document through her valuable typesetting advice.

This research was generously supported through a VIGRE grant of the National Science Foundation. The Mathematics Department of the University of Michigan and the Horace H. Rackham Graduate School also provided assistance.

Special thanks go to my wife Eva Ackermann and to my parents.

CHAPTER 1

# Introduction

The purpose of this paper is to work out the categorical basis for the foundations of conformal field theory. The definition of conformal field theory was outlined in Segal [**45**] and recently given in [**25**] and [**26**]. Concepts of 2-category theory, such as versions of algebra, limit, colimit, and adjunction, are necessary for this definition.

The structure present on the class $\mathcal{C}$ of rigged surfaces is captured by these concepts of 2-category theory. Here a *rigged surface* is a real, compact, not necessarily connected, two dimensional manifold with complex structure and analytically parametrized boundary components. Isomorphisms of such rigged surfaces are holomorphic diffeomorphisms preserving the boundary parametrizations. These rigged surfaces and isomorphisms form a groupoid and are part of the structure present on $\mathcal{C}$. Concepts of 2-categories enter when we describe the operations of disjoint union of two rigged surfaces and gluing along boundary components of opposite orientation. We need a mathematical structure to capture all of these features. This has been done in [**25**].

One step in this direction is the notion of algebra over a theory in the sense of Lawvere [**34**]. We need a weakened notion in which relations are replaced by coherence isos. This weakened notion is called a *pseudo algebra* in this paper. Coherence diagrams are required in a pseudo algebra, but it was noticed in [**25**] that Lawvere's notion of a theory allows us to write down all such diagrams easily. See Chapter 7 below. A symmetric monoidal category as defined in [**39**] provides us with a classical example of a pseudo algebra over the theory of commutative monoids. Theories, duality, and related topics are discussed further in [**1**], [**2**], [**3**], [**35**], and [**36**].

Unfortunately, pseudo algebras over a theory are not enough to capture the structure on $\mathcal{C}$. The reason is that the operation of gluing is indexed by the variable set of pairs of boundary components of opposite orientation. The operation of disjoint union also has an indexing. We need pseudo algebras over a "theory indexed over another theory," which we call a 2-theory. More precisely, the pseudo algebras we need are pseudo algebras over the 2-theory of *commutative monoids with cancellation.* See [**25**] and Chapter 13 below. The term 2-theory does *not* mean a theory in 2-categories.

Nevertheless, 2-categories are relevant. This is because we want to capture the behavior of holomorphic families of rigged surfaces in our description of the structure of $\mathcal{C}$. This amounts to saying that $\mathcal{C}$ is a stack of pseudo commutative monoids with cancellation. To consider this, we must remark that pseudo algebras over a theory and pseudo algebras over a 2-theory form 2-categories. A *stack* is a contravariant pseudo functor from a Grothendieck site into a 2-category which takes Grothendieck covers into limits of certain type, which are called bilimits. They are defined below, in [**29**], and [**50**], while a slightly stronger notion is called pseudo limit in [**50**]. One needs to understand such notions for the rigorous foundations of conformal field theory. More elaborate notions, such as analogous kinds of colimits are also needed in [**26**].

In this article we introduce the general concepts of weighted bilimits, weighted bicolimits, and biadjoints for pseudo functors between 2-categories in the sense below and prove statements about their existence in certain cases. There are many versions of such concepts and many (but not all) of the theorems we give are in the literature, see [**8**], [**11**], [**14**], [**19**], [**22**], [**21**], [**23**], [**31**], [**28**], [**29**], [**46**], [**48**], [**49**], [**50**], and [**51**]. Bicategories were first introduced in [**6**] and [**16**]. The circumstances of conformal field theory suggest a particular choice of concepts. To a topologist, the most natural and naive choice of terminology may be to use the term "lax" to mean "up to coherence isos" with these coherence isos required to satisfy appropriate coherence diagrams. "Iso" seems to be the only natural concept in the case of pseudo algebras over a theory: there seems to be no reasonable notion where coherences would not be iso. For this reason, the authors of [**25**], [**26**], and [**27**] use the "lax=up to coherence isos" philosophy. This terminology however turns out to be incorrect from the point of view of category theory (other ad hoc terminology also appears in [**25**], [**26**], and [**27**]). In this paper, we decided to follow established categorical terminology while giving a precise translation of the notions in [**25**], [**26**], and [**27**]. In the established categorical terminology, what is called a lax algebra in [**25**], [**26**], and [**27**] is called a *pseudo algebra*, what is called a lax morphism (morphism which commutes with operations up to coherence isos) in [**25**], [**26**], and [**27**] is called a *pseudo morphism* (or just *morphism*), and what is called a lax functor in [**25**], [**26**], and [**27**] is called a *pseudo functor*. In addition, the notions which the authors of [**25**], [**26**], and [**27**] refer to as lax limit, lax colimit, and lax adjoint are called *bilimit*, *bicolimit*, and *biadjoint* in established categorical terminology. The stronger categorical notions of pseudo limit, pseudo colimit, and pseudo adjoint are also sometimes relevant.

The term "lax" in standard categorical terminology is reserved for notions "up to 2-cells which are not necessarily iso". However, such notions will not play a central role in the present paper, as our motivation here is the same as in [**25**], [**26**], and [**27**], namely conformal field theory and stacks.

We show that every pseudo functor from a 1-category to the 2-category of small categories admits both a pseudo limit and a pseudo colimit by constructive proofs. Furthermore, the 2-category of small categories admits weighted pseudo limits and weighted pseudo colimits. After that we introduce the notions of a theory, an algebra over a theory, and a pseudo algebra over a theory. We then go on to show that any pseudo functor from a 1-category to the 2-category of pseudo $T$-algebras admits a pseudo limit by an adaptation of the proof for small categories. After a proof of the existence of cotensor products in the 2-category of pseudo $T$-algebras, we conclude from a theorem of Street that this 2-category admits weighted pseudo limits.

We continue the study of weakened structures by turning to biadjoints. First we show that a pseudo functor admits a left biadjoint if and only if for each object of the source category we have an appropriate biuniversal arrow in analogy to the standard result in 1-category theory. By means of this description we show that for any morphism of theories $\phi : S \to T$ the associated forgetful 2-functor from the 2-category of pseudo $T$-algebras to the 2-category of pseudo $S$-algebras admits a left biadjoint. The formalism developed for biadjoints is then adapted to treat bicolimits of pseudo $T$-algebras. Moreover, the universal property of these bicolimits is slightly weaker than the universal property of the pseudo limits. Similarly, the 2-category of pseudo $T$-algebras admits bitensor products, and hence also weighted bicolimits.

Lastly, we construct pseudo limits of pseudo algebras over a 2-theory. Again, a theorem of Street and the existence of cotensor products imply that the 2-category of pseudo algebras over a 2-theory admits weighted pseudo limits. An example of a pseudo algebra over a 2-theory comes from the category of rigged surfaces in [**25**].

Some of these results may be found in some form in the literature. There are many different ways to weaken 1-categorical concepts. This study only sets up the weakened notions needed for utilizing stacks to rigorously define conformal field theory as in [**25**]. Results about bilimits can be found in the references mentioned above. In particular, Gray explicitly describes quasilimits and quasicolimits of *strict* 2-functors from an arbitrary small 2-category to the 2-category $Cat$ of small categories on pages 201 and 219 of [**19**], although his quasilimit is defined in terms of quasiadjunction rather than cones. In any case, he does not have formulas for pseudo limits of *pseudo* functors. Street has the most general result in this context. In [**50**], he states that $Cat$ admits indexed pseudo limits of pseudo functors and writes down the indexed pseudo limit. His indexed pseudo limit is the same as the weighted pseudo limit in this paper. Results about notions similar to the notion of biadjoint can be found in [**19**], [**20**], [**29**], and [**50**]. These similarities are discussed in the introduction to Chapter 9. Blackwell, Kelly, and Power have limit and adjoint results similar to ours for strict 2-functors into 2-categories of strict algebras and pseudo morphisms over a 2-monad in [**9**]. In fact, we prove below that pseudo algebras over a theory are the strict algebras for a 2-monad in Chapter 7.

Any discussion of weakened algebraic structures must involve coherence questions, one of which was first treated in the classic paper [**37**] of Mac Lane. Many authors, including Laplaza, Kelly, Mac Lane, and Paré, have contributed to the theory of coherence as evidenced by the bibliographies of [**38**] and [**39**]. Some recent treatments in the context of $n$-categories and categorification are [**4**], [**5**], and [**15**]. See also [**53**], [**54**], and [**55**] for an approach to coherence involving a notion of 2-theory distinct from the notion of 2-theory in [**25**], [**26**], [**27**], and Chapter 13.

We follow the usual convention that 2-categories are denoted by capital script letters $\mathcal{A}, \mathcal{C}, \mathcal{D}, \mathcal{X}$, pseudo functors are denoted by capital letters $F, G$, morphisms are denoted by $e, f, g, h$, and 2-cells are denoted by Greek letters $\alpha, \beta, \gamma$. The identity 2-cell on a morphism $f$ is denoted $i_f$. Natural transformations and pseudo natural transformations are also denoted by lowercase Greek letters. The double arrow $\Rightarrow$ is used to denote 2-cells, natural transformations, and pseudo natural transformations, which in some cases are all the same thing. The notation $A \in \mathcal{D}$ means that $A$ is an object of $\mathcal{D}$.

We usually reserve the notation $\mathcal{C}$ for a 2-category in which we are building various limits and colimits. For example, in Chapters 4 and 5 the letter $\mathcal{C}$ denotes the 2-category of small categories, while it stands for the 2-category of small pseudo $T$-algebras in Chapters 8 and 11. In Chapter 13, the notation $\mathcal{C}$ stands for the 2-category of small pseudo $(\Theta, T)$-algebras. We use the same letter to highlight the similarities of the various proofs. In this introduction $\mathcal{C}$ stands for the category of rigged surfaces.

All sets, categories, and 2-categories appearing in this paper are assumed to be small.

CHAPTER 2

# Some Comments on Conformal Field Theory

In this chapter we make some motivational remarks about conformal field theory. Most of these terms will not appear in the rest of the paper, and are therefore only briefly discussed. More detail can be found in the articles [**25**] and [**26**], which this paper accompanies.

Conformal field theory has recently received considerable attention from mathematicians and physicists. It offers one approach to string theory, which aims to unify the four fundamental forces of nature. This is one reason why physicists are interested in conformal field theory as in [**43**]. The motivation for the axioms of conformal field theory comes from the path integral formalism of quantum field theory. Mathematicians have become interested in conformal field theory because it gives rise to a geometric definition of elliptic cohomology, which is related to Borcherds' proof [**12**] of the Moonshine conjectures.

The formalism necessary to rigorously define conformal field theory, and to prove theorems about it, is called *stacks of lax commutative monoids with cancellation (SLCMC's)* in [**25**]. These are the same as *stacks of pseudo algebras over the 2-theory of commutative monoids with cancellation* defined in Chapters 12 and 13. Roughly speaking, a strict commutative monoid with cancellation consists of a commutative monoid $I$ and a function $X : I^2 \longrightarrow Sets$ equipped with operations

$$+_{a,b,c,d} : X_{a,b} \times X_{c,d} \longrightarrow X_{a+c,b+d}$$

$$\check{?}_{a,b,c} : X_{a+c,b+c} \longrightarrow X_{a,b}$$

$$0 \in X_{0,0}$$

for all $a, b, c, d \in I$. These operations, called *disjoint union*, *cancellation (gluing)*, and *unit* must be commutative, associative, unital, and distributive in the appropriate senses. Whenever we add the adjective "pseudo" (or "lax" in [**25**], [**26**], [**27**]), it means that we replace sets by categories, functions by functors, and axioms by coherence isos that satisfy coherence diagrams. The theory and 2-theory apparatus gives us a concise way to list the necessary coherence isos and coherence diagrams. A thorough treatment of theories, 2-theories, their pseudo algebras, and their relevant diagrams are part of this paper. This formalism allows the authors of [**25**] and [**26**] to rigorously define conformal field theory in the sense of Segal, in particular all of the coherence isos and coherence diagrams are neatly encoded.

The first example of a pseudo commutative monoid with cancellation is the category of rigged surfaces. In this example the pseudo commutative monoid $I$ is the category of finite sets and bijections equipped with disjoint union. The 2-functor $X : I^2 \longrightarrow Cat$ from $I^2$ to the 2-category of small categories is given by defining $X_{a,b}$ to be the category of rigged surfaces with inbound components labelled by $a$ and outbound components labelled by $b$. The operation $+$ is disjoint union of labelled rigged surfaces (this is why the indices are added). The stack structure for this example is described in Section 13.3.

There are two other examples of SLCMC's that we need before defining conformal field theory and modular functor. These are $C(\mathcal{M})$ and $C(\mathcal{M}, H)$ from page 235 of [**26**]. The notation $\mathbb{C}_2$ denotes the pseudo commutative semi-ring of finite dimensional complex vector spaces, $\mathbb{C}_2^{Hilb}$ denotes the pseudo $\mathbb{C}_2$-algebra of complex Hilbert spaces equipped with the operation $\hat{\otimes}$ of Hilbert tensor product, $\mathcal{M}$ is a pseudo module over $\mathbb{C}_2$, $\mathcal{M}^{Hilb}$ denotes $\mathcal{M} \otimes \mathbb{C}_2^{Hilb}$, and $H$ is an object of $\mathcal{M}^{Hilb}$. If $\mathcal{M}$ has only one object, then $H$ is a Hilbert space, otherwise $H$ is a collection of Hilbert spaces indexed by the objects of $\mathcal{M}$. For finite sets $a, b$ the category $C(\mathcal{M})_{a,b}$ is $\mathcal{M}^{\otimes a} \otimes \mathcal{M}^{*\otimes b}$ where $\mathcal{M}^* := Hom_{pseudo}(\mathcal{M}, \mathbb{C}_2)$. The operation $+$ is given by $\otimes$ and gluing is given by evaluation $tr : \mathcal{M} \otimes \mathcal{M}^* \to \mathbb{C}_2$. The pseudo commutative monoid with cancellation $C(\mathcal{M}, H)$ is defined similarly, except that an object of $C(\mathcal{M}, H)_{a,b}$ is an object $M$ of $C(\mathcal{M})_{a,b}$ equipped with a morphism $M \to H^{\hat{\otimes} a} \hat{\otimes} H^{*\hat{\otimes} b}$ in $C(\mathcal{M})_{a,b}$ whose image consists of trace class elements. The morphisms of $C(\mathcal{M}, H)_{a,b}$ are the appropriate commutative triangles in $C(\mathcal{M})_{a,b}$. These two LCMC's can be made into stacks appropriately. Finally we are ready to give the rigorous definition of modular functor and conformal field theory.

DEFINITION 2.1. Let $C$ be a stack of pseudo commutative monoids with cancellation (SLCMC). A *modular functor on $C$ with labels $\mathcal{M}$* is a (pseudo) morphism $\phi : C \to C(\mathcal{M})$ of stacks of pseudo commutative monoids with cancellation. A *conformal field theory on $C$ with modular functor on labels $\mathcal{M}$ with state space $H$* is a (pseudo) morphism $\Phi : C \to C(\mathcal{M}, H)$ of stacks of pseudo commutative monoids with cancellation.

If we take $C$ to be the SLCMC of rigged surfaces, then we recover the usual definition of conformal field theory which assigns (up to a finite dimensional vector space) a trace class operator to a rigged surface in such a way that gluing surfaces corresponds to composing operators. Notice that modular functor and conformal field theory are both morphisms of the same algebraic structure. This was first noted by the authors of [**25**] and [**26**].

It is also possible to define one dimensional modular functors (*i.e.* those with one object in $\mathcal{M}$) in terms of $\mathbb{C}^\times$*-central extensions* of SLCMC's. A $\mathbb{C}^\times$-central extension of an SLCMC $\mathcal{D}$ is a strict morphism $\psi : \tilde{\mathcal{D}} \to \mathcal{D}$ of SLCMC's such that for fixed finite sets $s, t$, a fixed finite dimensional complex manifold $B$, and fixed $\alpha \in \mathcal{D}(B)_{s,t}$, the pre-images $\psi^{-1}(\alpha|_{B'})$ patch together for varying $B' \to B$ to form the sheaf of sections of a complex holomorphic line bundle over $B$. The maps on these sections induced by disjoint union and gluing are required to be isomorphisms of sheaves of vector spaces. If $\mathcal{H}$ is a Hilbert space, then there is an SLCMC $\underline{\mathcal{H}}$ in which $+$ is the operation of taking the Hilbert space tensor product and then the subset of trace class elements and $\check{?}$ is the trace map. Then a *chiral conformal field theory with one dimensional modular functor over $\mathcal{D}$* is a morphism of SLCMC's $\phi : \tilde{\mathcal{D}} \to \underline{\mathcal{H}}$ which is linear on the spaces of sections $\psi^{-1}(\alpha|_{B'})$.

The present paper deals with the 2-categorical foundations of the above project. We begin by introducing 2-categories and proving the existence of various types of limits in various 2-categories in Chapters 3, 4, 6, 8 , 11, and 13. We need the existence of certain limits in the above project because a stack is a contravariant pseudo functor that takes Grothendieck covers to bilimits. Grothendieck topologies and stacks are discussed in Chapter 12. The fundamentals of Lawvere theories and algebras are treated in Chapter 6. The passage from strict algebras to pseudo algebras, which is so important for the definition of conformal field theory, is discussed in Chapter 7. The biadjoints of Chapters 9 and 10 allow a universal description of the stack of covering spaces on page 337 of [**25**]. Lastly, the 2-theory of commutative monoids with cancellation is presented in Chapter 13 along with the example of rigged surfaces.

CHAPTER 3

# Weighted Pseudo Limits in a 2-Category

In this chapter we introduce the notion of a weighted pseudo limit and related concepts. The most important examples of 2-categories to keep in mind are the following.

EXAMPLE 3.1. The 2-category of small categories is formed by taking the objects (0-cells) to be small categories, the morphisms (1-cells) to be functors, and the 2-cells to be natural transformations. This 2-category is denoted *Cat*.

EXAMPLE 3.2. A full sub-2-category of the previous example is the 2-category with objects groupoids and 1-cells and 2-cells the same as above.

EXAMPLE 3.3. An example of a different sort is the 2-category with objects topological spaces, morphisms continuous maps, and 2-cells homotopy classes of homotopies. The 2-cells must be homotopy classes of homotopies in order to make the various compositions associative and unital.

EXAMPLE 3.4. Let $\mathcal{J}$ be a small 1-category. Then $\mathcal{J}$ has the structure of a 2-category if we regard $Mor_{\mathcal{J}}(i,j)$ as a discrete category for all $i,j \in Obj\ \mathcal{J}$.

These examples show that there are two ways of composing the 2-cells: vertically and horizontally. Natural transformations can be composed in two ways. Homotopy classes of homotopies can also be composed in two ways. To clarify which composition we mean, we follow Borceux's notation. See [**10**] for a more thorough discussion.

DEFINITION 3.5. Let $\mathcal{C}$ be a 2-category. If $A, B \in Obj\ \mathcal{C}$ and $f, g, h : A \to B$ are objects of the category $Mor(A, B)$ with 2-cells $\alpha : f \Rightarrow g$ and $\beta : g \Rightarrow h$ then the composition

$$\begin{array}{ccc} A & \xrightarrow{\quad f \quad} & B \\ & \Downarrow \alpha & \\ A & \xrightarrow{\quad g \quad} & B \\ & \Downarrow \beta & \\ A & \xrightarrow{\quad h \quad} & B \end{array}$$

in the category $Mor(A, B)$ is called the *vertical composition* of $\alpha$ and $\beta$. This composition is denoted $\beta \odot \alpha$. The *identity* on $f$ with respect to vertical composition is denoted $i_f$.

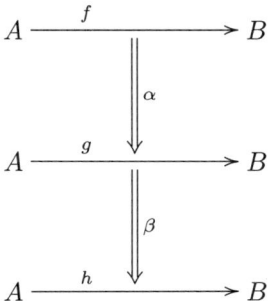

DEFINITION 3.6. Let $\mathcal{C}$ be a 2-category and $A, B, C \in Obj\ \mathcal{C}$. Let $c : Mor(A, B) \times Mor(B, C) \to Mor(A, C)$ denote the functor of composition in the 2-category $\mathcal{C}$. If $f, g : A \to B$ and $m, n : B \to C$ are objects of the respective categories $Mor(A, B)$ and $Mor(B, C)$ and $\alpha : f \Rightarrow g$, $\beta : m \Rightarrow n$ are 2-cells, then the composite 2-cell $c(\alpha, \beta) : c(f, m) \Rightarrow c(g, n)$ is called the *horizontal composition* of $\alpha$ and $\beta$. It is a morphism in the category $Mor(A, C)$ and is denoted $\beta * \alpha$.

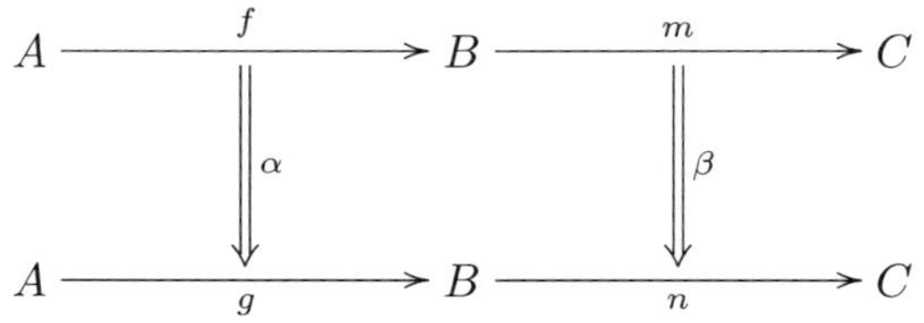


To define the concept of weighted pseudo limit, we need to discuss pseudo functors and pseudo natural transformations. A pseudo functor is like a 2-functor except that it preserves composition and identity only up to iso coherence 2-cells which satisfy coherence diagrams. A pseudo natural transformation is like a 2-natural transformation except that it is natural only up to an iso coherence 2-cell which satisfies coherence diagrams. We define these notions more carefully to fix some notation. We reproduce Borceux's treatment in [**10**]. The coherence 2-cells for pseudo functors and pseudo natural transformations in this paper are always assumed to be iso. Recall again that a pseudo functor in this paper is a lax functor in [**25**], [**26**], and [**27**] as well as in other previous papers.

DEFINITION 3.7. Let $\mathcal{C}, \mathcal{D}$ be 2-categories. A *pseudo functor* $F : \mathcal{C} \to \mathcal{D}$ consists of the following assignments and iso coherence 2-cells:

- For every object $A \in Obj\ \mathcal{C}$ an object $FA \in Obj\ \mathcal{D}$
- For all objects $A, B \in Obj\ \mathcal{C}$ a functor $F : Mor_{\mathcal{C}}(A, B) \to Mor_{\mathcal{D}}(FA, FB)$
- For all objects $A, B, C \in Obj\ \mathcal{C}$ a natural isomorphism $\gamma$ between the composed functors

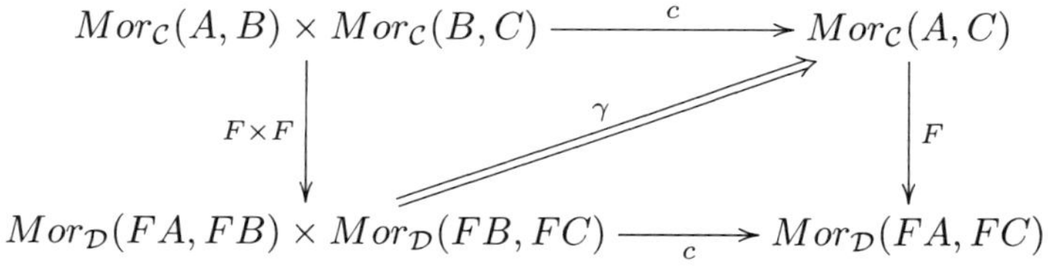


- For every object $A \in \mathcal{C}$ a natural isomorphism $\delta$ between the following composed functors.

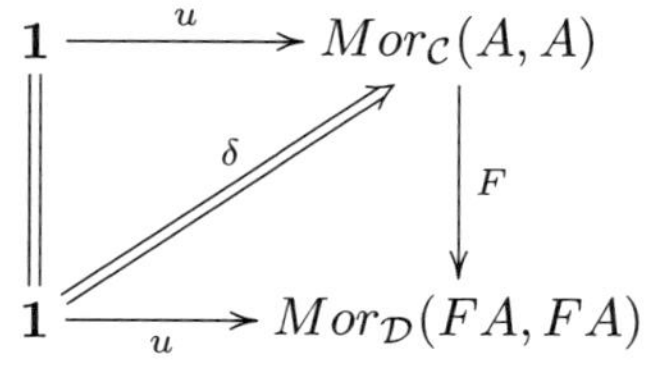


where the functor $u : \mathbf{1} \to Mor_{\mathcal{C}}(A, A)$ from the terminal object $\mathbf{1}$ in the category of small categories to the category $Mor_{\mathcal{C}}(A, A)$ takes the unique object $*$ of $\mathbf{1}$ to the identity morphism on $A$.

These coherence 2-cells must satisfy the following coherence diagrams.

- For every morphism $f : A \to B$ in $\mathcal{C}$ we require

$$\begin{array}{ccc} Ff \circ 1_{FA} & \xRightarrow{i_{Ff} * \delta_{A*}} & Ff \circ F1_A \\ {\scriptstyle i_{Ff}} \Downarrow & & \Downarrow {\scriptstyle \gamma_{1_A, f}} \\ Ff & \xRightarrow[i_{Ff}]{} & F(f \circ 1_A) \end{array} \qquad \begin{array}{ccc} 1_{FB} \circ Ff & \xRightarrow{\delta_{B*} * i_{Ff}} & F(1_B) \circ Ff \\ {\scriptstyle i_{Ff}} \Downarrow & & \Downarrow {\scriptstyle \gamma_{f, 1_B}} \\ Ff & \xRightarrow[i_{Ff}]{} & F(1_B \circ f) \end{array}$$

  to commute. Here $\delta_{A*}$ means the natural transformation $\delta_A$ evaluated at the unique object $*$ of $\mathbf{1}$. This is called the *unit axiom* for the pseudo functor $F$.
- For all morphisms $f, g, h$ of $\mathcal{C}$ such that $h \circ g \circ f$ exists we require that

$$\begin{array}{ccc} Fh \circ Fg \circ Ff & \xRightarrow{i_{Fh} * \gamma_{f,g}} & Fh \circ F(g \circ f) \\ {\scriptstyle \gamma_{g,h} * i_{Ff}} \Downarrow & & \Downarrow {\scriptstyle \gamma_{g \circ f, h}} \\ F(h \circ g) \circ Ff & \xRightarrow[\gamma_{f, h \circ g}]{} & F(h \circ g \circ f) \end{array}$$

  commutes. This is called the *composition axiom* for the pseudo functor $F$.

Each of these functors and natural transformations of course depends on the objects, so they really need indices, *e.g.* $c_{A,B,C}, F_{A,B}, \gamma_{A,B,C}, u_A, u_{FA}$, and $\delta_A$. Often we leave the indices off for more convenient notation. Note that the first diagram in the definition says that the pseudo functor preserves composition of morphisms up to coherence 2-cell because for morphisms $A \xrightarrow{f} B \xrightarrow{g} C$ in $\mathcal{C}$ we have $\gamma_{f,g} : F(g) \circ F(f) \Rightarrow F(g \circ f)$ and $\gamma$ is natural in $f$ and $g$. The second diagram in the definition says that the pseudo functor preserves identity up to coherence 2-cell because $\delta_{A*} : 1_{FA} \Rightarrow F(1_A)$ for all $A \in Obj\ \mathcal{C}$.

DEFINITION 3.8. Let $\mathcal{C} \xrightarrow{F} \mathcal{D} \xrightarrow{G} \mathcal{E}$ be pseudo functors. Then the *composition $G \circ F$ of pseudo functors* is the composition of the underlying maps of objects and the composition of the underlying functors on the morphism categories. The coherence 2-cells are as follows.

- For morphisms $A \xrightarrow{f} B \xrightarrow{g} C$ in $\mathcal{C}$ the 2-cell $\gamma^{GF}_{f,g}$ is the composition

$$GF(g) \circ GF(f) \xRightarrow{\gamma^G_{Ff,Fg}} G(Fg \circ Ff) \xRightarrow{G(\gamma^F_{f,g})} GF(g \circ f).$$

- For each object $A \in Obj\ \mathcal{C}$ the 2-cell $\delta^{GF}_{A*}$ is the composition

$$1_{GFA} \xRightarrow{\delta^G_{FA*}} G(1_{FA}) \xRightarrow{G(\delta^F_{A*})} GF(1_A).$$

Then the assignment $(f,g) \mapsto \gamma^{GF}_{f,g}$ is natural and $\gamma^{GF}$ and $\delta^{GF}_A$ satisfy the coherences to make $GF$ a pseudo functor.

DEFINITION 3.9. A *pseudo natural transformation* $\alpha : F \Rightarrow G$ from the pseudo functor $F : \mathcal{C} \to \mathcal{D}$ to the pseudo functor $G : \mathcal{C} \to \mathcal{D}$ consists of the following assignments:

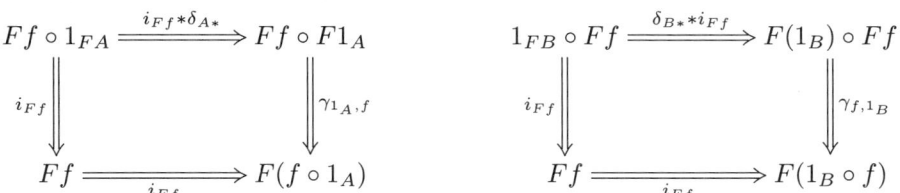

- For each $A \in Obj\ \mathcal{C}$ a morphism $\alpha_A : FA \to GA$ in the category $\mathcal{D}$
- For all objects $A, B \in Obj\ \mathcal{C}$ a natural isomorphism $\tau$ between the following functors.

$$\begin{array}{ccc} Mor_{\mathcal{C}}(A,B) & \xrightarrow{\quad F \quad} & Mor_{\mathcal{D}}(FA,FB) \\ {\scriptstyle G}\big\downarrow & \overset{\tau}{\nearrow\!\!\!\nearrow} & \big\downarrow{\scriptstyle \alpha_B \circ} \\ Mor_{\mathcal{D}}(GA,GB) & \xrightarrow[\circ \alpha_A]{\quad} & Mor_{\mathcal{D}}(FA,GB) \end{array}$$

The natural transformations $\tau$ must satisfy the following coherence diagrams involving $\delta$ and $\gamma$.

- For every $A \in Obj\ \mathcal{C}$ we require

$$\begin{array}{ccccc} \alpha_A & \xRightarrow{\quad i_{\alpha_A} \quad} & 1_{GA} \circ \alpha_A & \xRightarrow{\quad \delta^G_{A*} * i_{\alpha_A} \quad} & G(1_A) \circ \alpha_A \\ {\scriptstyle i_{\alpha_A}}\Big\Downarrow & & & & \Big\Downarrow{\scriptstyle \tau_{1_A}} \\ \alpha_A \circ 1_{FA} & & \xRightarrow[\quad i_{\alpha_A} * \delta^F_{A*} \quad]{} & & \alpha_A \circ F(1_A) \end{array}$$

  to commute. This is called the *unit axiom* for the pseudo natural transformation $\alpha$.

- For all morphisms $A \xrightarrow{f} B \xrightarrow{g} C$ in $\mathcal{C}$ we require

$$\begin{array}{ccccc} Gg \circ Gf \circ \alpha_A & \xRightarrow{\quad i_{Gg} * \tau_f \quad} & Gg \circ \alpha_B \circ Ff & \xRightarrow{\quad \tau_g * i_{Ff} \quad} & \alpha_C \circ Fg \circ Ff \\ {\scriptstyle \gamma^G_{f,g} * i_{\alpha_A}}\Big\Downarrow & & & & \Big\Downarrow{\scriptstyle i_{\alpha_C} * \gamma^F_{f,g}} \\ G(g \circ f) \circ \alpha_A & & \xRightarrow[\quad \tau_{g \circ f} \quad]{} & & \alpha_C \circ F(g \circ f) \end{array}$$

  to commute. This is called the *composition axiom* for the pseudo natural transformation $\alpha$.

Here $\tau$ should of course also be indexed by the objects $A, B$ *etc.*, but we leave off these indices for convenience. The coherence required on $\gamma$ and $\tau$ is the commutivity of the 2-cells (from $\tau$ and $\gamma$) written on the faces of the prism with edges $Ff$, $Fg$, $F(g \circ f)$, $Gf$, $Gg$, and $G(g \circ f)$ where $f$ and $g$ are composable morphisms in the 2-category $\mathcal{C}$. There are several ways to compose these 2-cells, but they are related by the interchange law. Here we must sometimes horizontally precompose or postcompose a 2-cell with identity 2-cells in order to horizontally compose. Note the diagram for $\tau$ drawn in the definition says that the assignment of $A \mapsto \alpha_A$ is natural up to coherence 2-cell because for $f \in Mor_{\mathcal{C}}(A, B)$ we have the diagram

$$\begin{array}{ccc} FA & \xrightarrow{\quad \alpha_A \quad} & GA \\ {\scriptstyle Ff}\big\downarrow & \overset{\tau_f}{\swarrow\!\!\!\swarrow} & \big\downarrow{\scriptstyle Gf} \\ FB & \xrightarrow[\alpha_B]{\quad} & GB \end{array}$$

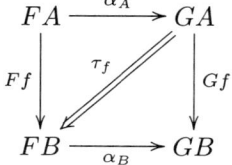

in $\mathcal{D}$. The assignment $f \mapsto \tau_f$ is natural in $f$, *i.e.* $\tau_{A,B}$ is a natural transformation.

Some authors prefer to denote the coherence 2-cells of $\alpha$ by $\alpha_f$ instead of $\tau_f$. However we follow Borceux's notation in [**10**] and use the distinguished notation $\tau$ in order to navigate complicated diagrams with less effort.

Pseudo natural transformations can also be horizontally and vertically composed. For example, if $F \overset{\alpha}{\Longrightarrow} G \overset{\beta}{\Longrightarrow} H$ are pseudo natural transformations, the vertical composition $\beta \odot \alpha$ has coherence 2-cells $\tau_f^{\beta\odot\alpha} = (i_{\beta_B} * \tau_f^\alpha) \odot (\tau_f^\beta * i_{\alpha_A})$ for $f : A \to B$ as in the following diagram.

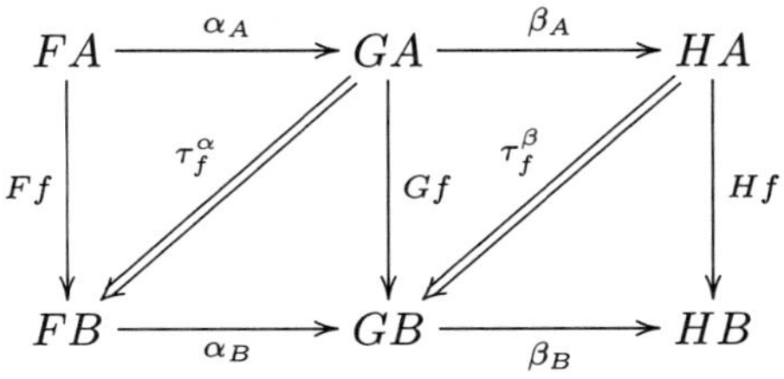


Natural transformations can be seen as morphisms between functors. In the context of 2-categories there is a similar notion of a modification between pseudo natural transformations.

DEFINITION 3.10. Let $F, G : \mathcal{C} \to \mathcal{D}$ be pseudo functors and $\alpha, \beta : F \Rightarrow G$ pseudo natural transformations. A *modification* $\Xi : \alpha \rightsquigarrow \beta$ is a function which assigns to every $A \in \mathit{Obj}\ \mathcal{C}$ a 2-cell $\Xi_A : \alpha_A \Rightarrow \beta_A$ in $\mathcal{D}$ in such a way that $\tau^\beta_{A,B}(g) \odot (G\gamma * \Xi_A) = (\Xi_B * F\gamma) \odot \tau^\alpha_{A,B}(f)$ for all $A, B \in \mathit{Obj}\ \mathcal{C}$ and all morphisms $f, g : A \to B$ and all 2-cells $\gamma : f \Rightarrow g$. Here $\tau^\alpha$ and $\tau^\beta$ denote the natural transformations belonging to the pseudo natural transformations $\alpha$ and $\beta$ respectively, while $\gamma$ is an arbitrary 2-cell in $\mathcal{C}$. This means that the following two compositions of 2-cells are the same.

(3.1)

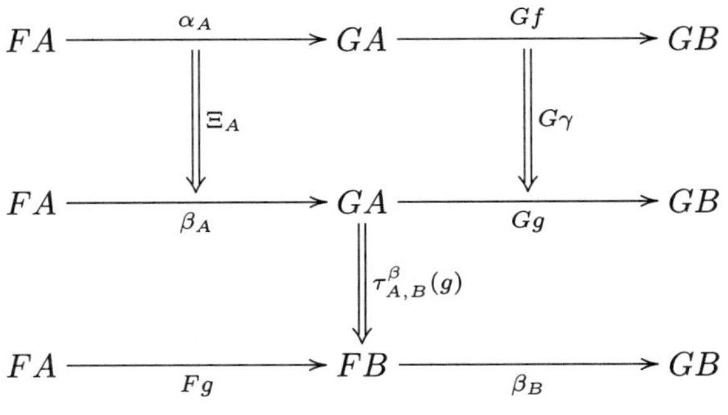

$$
\begin{array}{ccccc}
FA & \xrightarrow{\alpha_A} & GA & \xrightarrow{Gf} & GB \\
 & & \Big\Downarrow \tau^{\alpha}_{A,B}(f) & & \\
FA & \xrightarrow{Ff} & FB & \xrightarrow{\alpha_B} & GB \\
 & \Big\Downarrow F\gamma & & \Big\Downarrow \Xi_B & \\
FA & \xrightarrow[Fg]{} & FB & \xrightarrow[\beta_B]{} & GB
\end{array}
\tag{3.2}
$$

These two diagrams can be combined to make a cube whose faces have 2-cells inscribed in them. In this definition $\gamma$ is not to be confused with the required coherence 2-cell in the definition of pseudo functor.

DEFINITION 3.11. If $F : \mathcal{D} \to \mathcal{C}$ is a pseudo functor, then a *pseudo limit* of $F$ consists of an object $W \in Obj\ \mathcal{C}$ and a pseudo natural transformation $\pi : \Delta_W \Rightarrow F$ from the constant 2-functor $W$ to the pseudo functor $F$ which is universal in the following sense: the functor $(\pi\circ) : Mor_{\mathcal{C}}(C, W) \to PseudoCone(C, F)$ is an *isomorphism* of categories for every object $C \in Obj\ \mathcal{C}$.

$PseudoCone(C, F)$ denotes here the category with objects taken to be the pseudo natural transformations $\Delta_C \Rightarrow F$ and with morphisms taken to be the modifications. Pseudo colimits can be defined in terms of $PseudoCone(F, C)$ and $(\circ\pi) : Mor_{\mathcal{C}}(W, C) \to PseudoCone(F, C)$ similarly.

THEOREM 3.12. *Any two pseudo limits of a pseudo functor are isomorphic.*

DEFINITION 3.13. If $F : \mathcal{D} \to \mathcal{C}$ is a pseudo functor, then a *bilimit* of $F$ consists of an object $W \in Obj\ \mathcal{C}$ and a pseudo natural transformation $\pi : \Delta_W \Rightarrow F$ from the constant 2-functor $W$ to the pseudo functor $F$ which is universal in the following sense: the functor $(\pi\circ) : Mor_{\mathcal{C}}(C, W) \to PseudoCone(C, F)$ is an *equivalence* of categories for every object $C \in Obj\ \mathcal{C}$.

Some authors would call this bilimit a *conical bilimit*, see [**29**] and [**50**] for example. They discuss the more general notion of *weighted bilimit* or *indexed bilimit*, which is defined below. Limits defined in terms of cones, such as this bilimit, have constant weight or constant index. For our applications to conformal field theory, it is sufficient to consider only conical bilimits although we prove results for more general weighted bilimits in this paper. The existence of conical bilimits is sufficient to speak of stacks. The term *lax limit* in [**25**], [**26**], and [**27**] is synonymous with the term *bilimit* defined above.

Every pseudo limit for a fixed pseudo functor is obviously a bilimit of that pseudo functor. One can ask whether or not bilimits and pseudo limits are the same. The following trivial example shows that bilimits and pseudo limits are not the same.

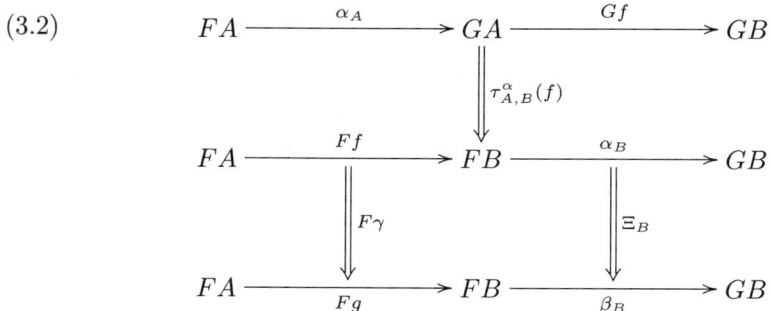

EXAMPLE 3.14. Let $\mathbf{1}$ denote the terminal object in the category of small categories, in other words $\mathbf{1}$ is the category with one object $*$ and one morphism, namely the identity morphism. This category can be viewed as a 2-category with no nontrivial 2-cells. Suppose $\mathcal{C}$ is a 2-category with at least two objects $W, W'$ such that we have a morphism $\pi' : W' \to W$ which is a pseudo isomorphism. This

means that there exists a morphism $\theta : W \to W'$ and iso 2-cells $\theta \circ \pi' \Rightarrow 1_{W'}$ and $\pi' \circ \theta \Rightarrow 1_W$. Suppose further that $\pi'$ is not monic. This means there exists an object $C \in Obj\, \mathcal{C}$ and distinct morphisms $f_1, f_2 : C \to W'$ such that $\pi' \circ f_1 = \pi' \circ f_2$. Let $F : \mathbf{1} \to \mathcal{C}$ be the constant functor $\Delta_W$, *i.e.* $F(*) = W$ and the identity gets mapped to $1_W$. Then $PseudoCone(C, F)$ is isomorphic to $Mor_{\mathcal{C}}(C, W)$. We identify these two categories. Obviously $W$ and the pseudo natural transformation $\pi = 1_W$ (under the identification) form a pseudo limit, while $W'$ and $\pi'$ form a bilimit. However, $W'$ and $\pi'$ do not form a pseudo limit because $(\pi' \circ) : Mor_{\mathcal{C}}(C, W') \to Mor_{\mathcal{C}}(C, W)$ is not an isomorphism of categories, since $\pi' \circ f_1 = \pi' \circ f_2$ although $f_1 \neq f_2$.

EXAMPLE 3.15. There are also examples where a bicolimit exists but not a pseudo colimit. This example goes back to **[9]**. Let *Lex* denote the 2-category of small finitely complete categories, left exact functors, and natural transformations. A functor is called *left exact* if it preserves all finite limits. An initial object is a colimit of the empty 2-functor. A pseudo colimit and a 2-colimit of the empty 2-functor are the same thing. The 2-category *Lex* does not admit an initial object because there are always two distinct functors $A \to I$ where $I$ is the category with only two isomorphic objects and no nontrivial morphisms besides the isomorphism and its inverse. The two constant functors provide us with two distinct functors $A \to I$ for each $A \in Obj\, Lex$. The empty functor does however admit a bicolimit because *Lex* is the 2-category of strict algebras, pseudo algebra morphisms, and 2-cells for some finitary 2-monad on *Cat*. Blackwell, Kelly, and Power prove in **[9]** that such algebra categories admit bicolimits.

Many pseudo algebra categories do not admit pseudo colimits because the morphisms are not strict. Another example can be obtained by adapting Example 10.14 on page 126 to colimits.

After Example 3.14, one might wonder whether or not the equivalences of categories in the definition of bilimit can be chosen in some natural way. They can in fact be chosen pseudo naturally as follows. We write it explicitly only for the bicolimit, although a completely analogous statement holds for the bilimit.

REMARK 3.16. Let $\mathcal{C}, \mathcal{D}$ be 2-categories. Let $\mathcal{F} : \mathcal{D} \to \mathcal{C}$ be a pseudo functor. Suppose $W \in Obj\, \mathcal{C}$ is a bicolimit with universal pseudo cone $\pi : \mathcal{F} \Rightarrow \Delta_W$. Let $\phi_C$ denote the equivalence of categories $(\circ\pi) : Mor_{\mathcal{C}}(W, C) \to PseudoCone(\mathcal{F}, C)$. Let $G(C) := Mor_{\mathcal{C}}(W, C)$ and $F(C) := PseudoCone(\mathcal{F}, C)$. Then $G$ and $F$ are strict 2-functors and $C \mapsto \phi_C$ is a 2-natural transformation $G \Rightarrow F$.

*Proof:* This follows from the definitions. □

REMARK 3.17. Let the notation be the same as in the previous remark. For $C \in Obj\, \mathcal{C}$ let $\psi_C : FC \to GC$ be a right adjoint to $\phi_C$ such that the unit $\eta_C : 1_{GC} \Rightarrow \psi_C \circ \phi_C$ and counit $\varepsilon_C : \phi_C \circ \psi_C \Rightarrow 1_{FC}$ are natural isomorphisms. Then $C \mapsto \psi_C$ is a pseudo natural transformation from $F$ to $G$ and there exist iso modifications $\eta : i_G \rightsquigarrow \psi \odot \phi$ and $\varepsilon : \phi \odot \psi \rightsquigarrow i_F$ which satisfy the triangle identities, namely $C \mapsto \eta_C$ and $C \mapsto \varepsilon_C$. In the terminology of **[50]**, this means that $F$ and $G$ are equivalent in the 2-category $Hom[\mathcal{C}, Cat]$ of pseudo functors, pseudo natural transformations, and modifications. The equivalences in $Hom[\mathcal{C}, Cat]$ are precisely the pseudo natural transformations whose components are equivalences of categories.

*Proof:* Since $\phi_C$ is an equivalence of categories, there exists such a functor $\psi_C$ with unit and counit as above. For $f : A \to B$ in $\mathcal{C}$ define the coherence iso $\tau_f^\psi : Gf \circ \psi_A \Rightarrow \psi_B \circ Ff$ to be the composition of 2-cells in the following diagram.

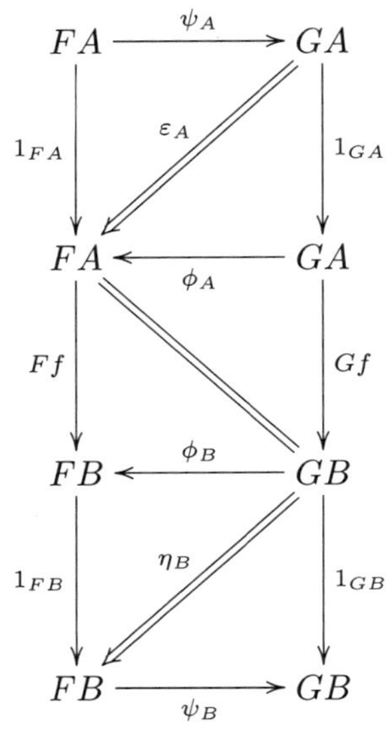


The middle square commutes because $\phi$ is a 2-natural transformation. We can see that the assignment $f \mapsto \tau_f^\psi$ is natural after segmenting the naturality diagram into three inner squares and using the fact that $\phi$ is a 2-natural transformation as follows. Let $f, g : A \to B$ and $\mu : f \to g$ in $\mathcal{C}$.

$$\begin{array}{ccccccc}
1_{GB} \circ Gf \circ \psi_A & \xRightarrow{\eta_B * i_{Gf} * i_{\psi_A}} & \psi_B \circ \phi_B \circ Gf \circ \psi_A & = & \psi_B \circ Ff \circ \phi_A \circ \psi_A & \xRightarrow{i_{\psi_B} * i_{Ff} * \varepsilon_A} & \psi_B \circ Ff \circ 1_{FA} \\
\Big\Downarrow {\scriptstyle i_{1_{GB}} * G\mu * i_{\psi_A}} & & \Big\Downarrow {\scriptstyle i_{\psi_B \circ \phi_B} * G\mu * i_{\psi_A}} & & \Big\Downarrow {\scriptstyle i_{\psi_B} * F\mu * i_{\phi_A \circ \psi_A}} & & \Big\Downarrow {\scriptstyle i_{\psi_B} * F\mu * i_{1_{FA}}} \\
1_{GB} \circ Gg \circ \psi_A & \xRightarrow[\eta_B * i_{Gg} * i_{\psi_A}]{} & \psi_B \circ \phi_B \circ Gg \circ \psi_A & = & \psi_B \circ Fg \circ \phi_A \circ \psi_A & \xRightarrow[i_{\psi_B} * i_{Fg} * \varepsilon_A]{} & \psi_B \circ Fg \circ 1_{FA}
\end{array}$$

The left square and the right square commute because of the interchange law and the defining property of identity 2-cells. The middle square commutes because $\phi$ is a 2-natural transformation. Hence the outermost rectangle commutes and $f \mapsto \tau_f^\psi$ is natural.

Since $F$ and $G$ are strict 2-functors, verifying the unit axiom for $\psi$ reduces to proving that $\tau_{1_C}^\psi$ is $i_{\psi_C}$ for all $C \in Obj\ \mathcal{C}$. That follows from the definition of $\tau_{1_C}$ and one of the triangle identities.

Since $F$ and $G$ are strict 2-functors, verifying the composition axiom for $\psi$ amounts to proving for $A \xrightarrow{f} B \xrightarrow{g} C$ in $\mathcal{C}$ that the composition $(\tau_g^\psi * i_{Ff}) \odot (i_{Gg} * \tau_f^\psi)$ in (3.3) is the same as $\tau_{g \circ f}^\psi$ in (3.4). That follows since the middle parallelogram in (3.4) is $i_{\phi_B}$ by the triangle identity. Hence $\psi$ with $\tau^\psi$ satisfies the composition axiom and we conclude that $C \mapsto \psi_C$ is a pseudo natural transformation $F \Rightarrow G$.

(3.3)

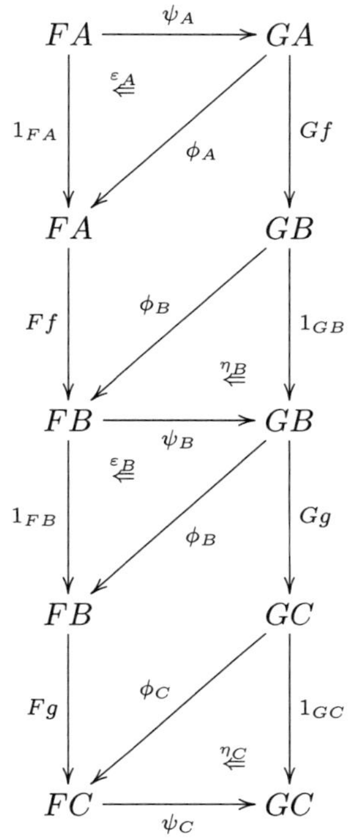


(3.4)

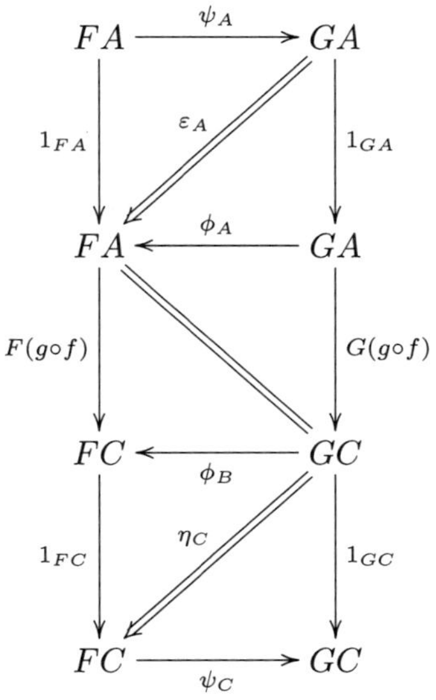


Next we prove that $A \mapsto \eta_A$ is a modification $i_G \rightsquigarrow \psi \odot \phi$. This requires a proof that (3.1) is the same as (3.2). Let $f, g : A \to B$ be morphisms in $\mathcal{C}$ and $\gamma : f \Rightarrow g$ a 2-cell in $\mathcal{C}$. Since $\phi$ is a 2-natural transformation, we see that (3.2) is $\eta_B * G\gamma$. We proceed by showing that (3.1) is $\eta_B * G\gamma$. Note that $\tau^{\beta}_{A,B}(g) = \tau_g^{\psi\circ\phi}$ in (3.1) is $(i_{\psi_B} * i_{\phi_B \circ Gg}) \odot (\tau_g^{\psi} * i_{\phi_A})$ by the remarks on page 13 about coherence isos for a vertical composition of pseudo natural transformations. Writing out (3.1) with $\alpha = i_G$, $\beta = \psi \odot \phi$, $\Xi = \eta$, and including many trivial arrows gives (3.5).

(3.5)

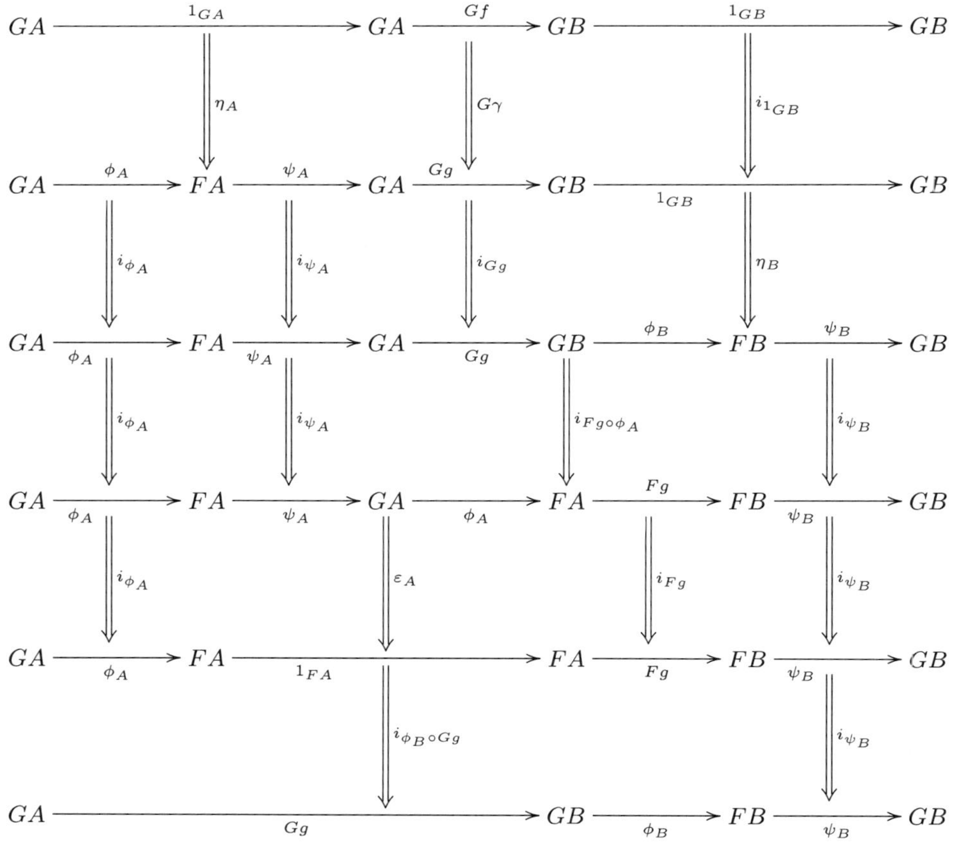


Using a triangle identity and contracting all the trivial identities, we see that the only thing that does not cancel is $\eta_B * G\gamma$. Hence (3.1) is the same as (3.2) and $A \mapsto \eta_A$ is a modification.

One can similarly show that $A \mapsto \varepsilon_A$ is a modification. The modifications $\eta$ and $\varepsilon$ satisfy the triangle identities because their constituent arrows do. □

DEFINITION 3.18. A 2-category $\mathcal{C}$ *admits bilimits* if every pseudo functor $F : \mathcal{J} \to \mathcal{C}$ from a small 1-category $\mathcal{J}$ to $\mathcal{C}$ admits a bilimit in $\mathcal{C}$.

There are analogous definitions for pseudo limits, bicolimits, and pseudo colimits. If we view the category $\mathcal{J}$ as an indexing category, then we can speak of bilimits of diagrams, *i.e.* we can view a diagram in $\mathcal{C}$ as the image of a pseudo functor from a source diagram $\mathcal{J}$ to the 2-category $\mathcal{C}$.

The concept of pseudo limit can be further generalized to weighted pseudo limit. For any small 2-category $\mathcal{C}$ we denote the small category $Mor_{\mathcal{C}}(A, B)$ by $\mathcal{C}(A, B)$ for $A, B \in Obj\ \mathcal{C}$.

DEFINITION 3.19. Let $\mathcal{C}, \mathcal{D}$ be 2-categories. Let $J : \mathcal{D} \to Cat$ and $F : \mathcal{D} \to \mathcal{C}$ be pseudo functors. Let $Hom[\mathcal{D}, Cat]$ denote the 2-category with pseudo functors

$\mathcal{D} \to Cat$ as objects, pseudo natural transformations as morphisms, and modifications as 2-cells. Then $\{J, F\}_p \in Obj\ \mathcal{C}$ is called a *J-weighted pseudo limit of F* if the strict 2-functors $\mathcal{C}^{op} \to Cat$

$$C \mapsto \mathcal{C}(C, \{J, F\}_p)$$

$$C \mapsto Hom[\mathcal{D}, Cat](J, \mathcal{C}(C, F-))$$

are 2-isomorphic. The image $\xi : J \Rightarrow \mathcal{C}(\{J, F\}_p, F-)$ of $1_{\{J,F\}_p}$ under this 2-representation is called the *unit.*

Street refers to this as the *J-indexed pseudo limit of F* in [**50**], although now the term weighted is used instead of indexed. This is similar to Kelly's definition in [**29**], except that his definition is for strict 2-functors $J, F$ and he uses the full sub-2-category $Psd[\mathcal{D}, Cat]$ of $Hom[\mathcal{D}, Cat]$ in place of $Hom[\mathcal{D}, Cat]$. The 2-category $Psd[\mathcal{D}, Cat]$ consists of strict 2-functors, pseudo natural transformations, and modifications.

We recover the usual definition of pseudo limit whenever $J$ is the constant functor which takes everything to the trivial category with one object. A weighted pseudo limit is said to be *conical* whenever $J$ is this constant functor. Another special type of weighted limit called *cotensor product* occurs when $\mathcal{D}$ is the trivial 2-category with one object and $J$ and $F$ are strict 2-functors. In this case $J$ and $F$ can be identified with objects of $Cat$ and $\mathcal{C}$ respectively. Tensor products can be defined similarly.

DEFINITION 3.20. Let $J \in Obj\ Cat$ and $F \in Obj\ \mathcal{C}$. Then $\{J, F\} \in Obj\ \mathcal{C}$ is called a *cotensor product* of $J$ and $F$ if the strict 2-functors $\mathcal{C}^{op} \to Cat$

$$C \mapsto \mathcal{C}(C, \{J, F\})$$

$$C \mapsto Cat(J, \mathcal{C}(C, F))$$

are 2-naturally isomorphic.

REMARK 3.21. (Kelly) We can rephrase the definition of cotensor product entirely in terms of the unit $\pi : J \to \mathcal{C}(\{J, F\}, F)$. The object $\{J, F\}$ of $\mathcal{C}$ is a cotensor product of $J$ and $F$ with unit $\pi : J \to \mathcal{C}(\{J, F\}, F)$ if and only if the functor $\mathcal{C}(C, \{J, F\}) \to Cat(J, \mathcal{C}(C, F))$ defined by composition with $\pi$

$$b \mapsto \mathcal{C}(b, F) \circ \pi$$

$$\alpha \mapsto \mathcal{C}(\alpha, F) * i_\pi$$

for arrows $b : C \to \{J, F\}$ and 2-cells $\alpha : b \to b'$ in $\mathcal{C}$ is an isomorphism of categories for all $C \in Obj\ \mathcal{C}$. More specifically:

(1) For every functor $\sigma : J \to \mathcal{C}(C, F)$ there is a unique arrow $b : C \to \{J, F\}$ in $\mathcal{C}$ such that $\mathcal{C}(b, F) \circ \pi = \sigma$.
(2) For every natural transformation $\Xi : \sigma \Rightarrow \sigma'$ there is a unique 2-cell $\alpha : b \Rightarrow b'$ in $\mathcal{C}$ such that $\mathcal{C}(\alpha, F) * i_\pi = \Xi$.

A useful reformulation of an observation by Street on page 120 of [**50**] illustrates the importance of cotensor products in the context of weighted pseudo limits.

THEOREM 3.22. *(Street) A 2-category $\mathcal{C}$ admits weighted pseudo limits if and only if it admits 2-products, cotensor products, and pseudo equalizers.*

REMARK 3.23. (Street) Pseudo equalizers can be constructed from cotensor products and 2-pullbacks, while 2-pullbacks can be constructed from 2-products and 2-equalizers. Thus it is sufficient to require 2-equalizers instead of pseudo equalizers in the previous theorem.

DEFINITION 3.24. Let $\mathcal{C}, \mathcal{D}$ be 2-categories. Let $J : \mathcal{D} \to Cat$ and $F : \mathcal{D} \to \mathcal{C}$ be pseudo functors. As above, let $Hom[\mathcal{D}, Cat]$ denote the 2-category with pseudo functors $\mathcal{D} \to Cat$ as objects, pseudo natural transformations as morphisms, and modifications as 2-cells. Then $\{J, F\}_b \in Obj\ \mathcal{C}$ is called a *J-weighted bilimit of F* if the strict 2-functors $\mathcal{C}^{op} \to Cat$

$$C \mapsto \mathcal{C}(C, \{J, F\}_b)$$
$$C \mapsto Hom[\mathcal{D}, Cat](J, \mathcal{C}(C, F-))$$

are equivalent in the 2-category $Hom[\mathcal{C}^{op}, Cat]$, *i.e.* there is a pseudo natural transformation going from one to the other whose arrow components are equivalences of categories. The image $\xi : J \Rightarrow \mathcal{C}(\{J, F\}_b, F-)$ of $1_{\{J,F\}_b}$ under this birepresentation is called the *unit*.

Kelly refers to this in [**29**] as the *J-indexed bilimit of F*. The concepts weighted bicolimit and bitensor product can be defined similarly. Later we will need bitensor products, so we formulate this precisely and describe it entirely in terms of the unit like Kelly in [**29**].

DEFINITION 3.25. Let $J \in Obj\ Cat$ and $F \in Obj\ \mathcal{C}$. Then $J * F \in Obj\ \mathcal{C}$ is called a *bitensor product* of $J$ and $F$ if the strict 2-functors $\mathcal{C}^{op} \to Cat$

$$C \mapsto \mathcal{C}(J * F, C)$$
$$C \mapsto Cat(J, \mathcal{C}(F, C))$$

are equivalent in the 2-category $Hom[\mathcal{C}^{op}, Cat]$.

REMARK 3.26. We can rephrase the definition of bitensor product entirely in terms of the unit $\pi : J \to \mathcal{C}(F, J * F)$. The object $J * F$ of $\mathcal{C}$ is a bitensor product of $J$ and $F$ with unit $\pi : J \to \mathcal{C}(F, J * F)$ if and only if the functor $\mathcal{C}(J * F, C) \to Cat(J, \mathcal{C}(F, C))$ defined by

$$b \mapsto \mathcal{C}(F, b) \circ \pi$$
$$\alpha \mapsto \mathcal{C}(F, \alpha) * i_\pi$$

for arrows $b : J * F \to C$ and 2-cells $\alpha : b \to b'$ in $\mathcal{C}$ is an equivalence of categories for all $C \in Obj\ \mathcal{C}$.

Street points out the dual version of the following theorem on page 120 of [**50**].

THEOREM 3.27. *A 2-category $\mathcal{C}$ admits weighted bicolimits if and only if it admits bicoproducts, bitensor products, and bicoequalizers.*

Cotensor products, bitensor products, and the theorems above will be used later to show that the 2-categories of interest to us admit weighted pseudo limits as well as weighted bicolimits.

CHAPTER 4

# Weighted Pseudo Colimits in the 2-Category of Small Categories

In this chapter we show constructively that the 2-category $\mathcal{C}$ of small categories admits pseudo colimits. The dual version of Theorem 3.22 will imply that this 2-category also admits weighted pseudo colimits. One of the concepts in the proof is the free category generated by a directed graph.

DEFINITION 4.1. A *directed graph* $G$ consists of a set $O$ of *objects* and a set $A$ of *arrows* and two functions $S, T : A \to O$ called *source* and *target.*

A directed graph is like a category except composition and identity arrows are not necessarily defined. Any directed graph $G$ whose sets of arrows and objects are both small generates a *free category* on $G$, which is also called the *path category* of $G$. Similarly $G$ generates a *free groupoid.* We can force commutivity of certain diagrams by putting a congruence on the morphism sets of the free category or free groupoid and then passing to the *quotient category.* We use this construction in the proof below. The $S, T$ in the definition of directed graph will also be used to denote the source and target of a morphism in a category.

THEOREM 4.2. *The 2-category $\mathcal{C}$ of small categories admits pseudo colimits.*

*Proof:* Let $\mathcal{J}$ be a small 1-category and $F : \mathcal{J} \to \mathcal{C}$ a pseudo functor. Here we view $\mathcal{J}$ as a 2-category which has no nontrivial 2-cells. The category $\mathcal{J}$ plays the role of an indexing category. For any $X \in Obj\ \mathcal{C}$ let $\Delta_X$ denote the constant 2-functor which takes every object of $\mathcal{J}$ to $X$, every morphism to $1_X$, and every 2-cell to the identity 2-cell $i_X : 1_X \Rightarrow 1_X$. Then a pseudo cone from $F$ to $X$ is a pseudo natural transformation $F \Rightarrow \Delta_X$. Recall $PseudoCone(F, X)$ denotes the category with objects the pseudo cones from $F$ to $X$ with morphisms the modifications between them. The pseudo colimit of $F$ is an object $W \in \mathcal{C}$ with a pseudo cone $\pi : F \Rightarrow \Delta_W$ which are universal in the sense that $(\circ\pi) : Mor_{\mathcal{C}}(W, V) \to PseudoCone(F, V)$ is an isomorphism of categories for all small categories $V$.

First we define candidates $W \in Obj\ \mathcal{C}$ and $\pi : F \Rightarrow \Delta_W$. Then we show that they are universal. For each $j \in Obj\ \mathcal{J}$ let $A_j$ denote the small category $Fj$ and let $a_f$ denote the functor $Ff$ between small categories. Since $F$ is a pseudo functor, for every pair $f, g$ of morphisms of $\mathcal{J}$ such that $g \circ f$ exists we have a natural transformation (a 2-cell in the 2-category of small categories) $\gamma_{f,g} : Fg \circ Ff \Rightarrow F(g \circ f)$. We define a directed graph with objects $O$ and arrows $A$ as follows. Let $O = \coprod_{j \in \mathcal{J}} Obj\ A_j$. There is a well defined function $p : O \to Obj\ \mathcal{J}$ satisfying $p(Obj\ A_j) = \{j\}$ because this union is disjoint, *i.e.* even if the small categories $A_i$ and $A_j$ are the same, we distinguish them in the disjoint union by their indices. Let the collection of arrows be $A = (\coprod_{j \in \mathcal{J}} Mor\ A_j) \coprod \{h_{(x,f)}, h^{-1}_{(x,f)} : (x, f) \in O \times Mor\ \mathcal{J}$ such that $p(x) = Sf\}$ where the elements of $\coprod_{j \in \mathcal{J}} Mor\ A_j$ have the obvious source and target while $Sh_{(x,f)} = x$ and $Th_{(x,f)} = a_f(x)$. Let $W'$ be the free category generated by this graph. We put the smallest congruence $\sim$ on $Mor\ W'$ such that:

- All of the relations in each $A_i$ are contained in $\sim$, *i.e.* for $m, n \in Mor\ A_i \subseteq Mor\ W'$ with $Sn = Tm$ we have $n \circ_{W'} m \sim n \circ_{A_i} m$ where the composition on the left is the composition in the free category $W'$ and the composition on the right is the composition in the small category $A_i$.
- For all $f, g \in Mor\ \mathcal{J}$ with $Sg = Tf$ and all $x \in Obj\ A_{Sf}$ we have $\gamma_{f,g}(x) \circ_{W'} h_{(a_f(x),g)} \circ_{W'} h_{(x,f)} \sim h_{(x,g\circ f)}$ and also every identity $1_x \in A_i$ is congruent to the identity in the free category on the object $x$.
- For all $i, j \in Obj\ \mathcal{J}$ and all $f \in Mor_{\mathcal{J}}(i,j)$ and all morphisms $m : x \to y$ of $A_i$ we have $h_{(y,f)} \circ_{W'} m \sim a_f(m) \circ_{W'} h_{(x,f)}$.
- For all $j \in Obj\ \mathcal{J}$ and all $x \in Obj\ A_j$ we have $(\delta^F_{j*})_x \sim h_{(x,1_j)}$ where $*$ denotes the unique object of the terminal object $\mathbf{1}$ in the category of small categories and $\delta^F_{j*}$ is the natural transformation $\delta^F_j$ evaluated at $*$.
- For all $h_{(x,f)}$ from above we have $h^{-1}_{(x,f)} \circ_{W'} h_{(x,f)} \sim 1_x$ and $h_{(x,f)} \circ_{W'} h^{-1}_{(x,f)} \sim 1_{a_f x}$.

Define $W$ to be the quotient category of the free category $W'$ by the congruence $\sim$. This is the candidate for the pseudo colimit.

Now we define a pseudo natural transformation $\pi : F \Rightarrow \Delta_W$ and its coherence 2-cells $\tau$, *i.e.* we define an element of $PseudoCone(F, W)$. For each object $j \in Obj\ \mathcal{J}$ we need a morphism in $\mathcal{C}$ (*i.e.* a functor) $\pi_j : Fj = A_j \to W = \Delta_W(j)$. Define $\pi_j : A_j \to W$ to be the inclusion functors $A_j \hookrightarrow W$. In order for $\pi$ to be a pseudo natural transformation, this assignment must be natural up to coherence 2-cell, *i.e.* for all $i, j \in Obj\ \mathcal{J}$ we should have a natural isomorphism $\tau_{i,j}$ of the following sort.

$$\begin{array}{ccc}
Mor_{\mathcal{J}}(i,j) & \xrightarrow{\quad F \quad} & Mor_{\mathcal{C}}(A_i, A_j) \\
{\scriptstyle \Delta_W}\downarrow & \overset{\tau_{i,j}}{\nearrow\!\!\!\!\nearrow} & \downarrow{\scriptstyle \pi_j \circ} \\
Mor_{\mathcal{C}}(W, W) & \xrightarrow[\circ \pi_i]{\qquad} & Mor_{\mathcal{C}}(A_i, W)
\end{array}$$

Evaluating this diagram at a morphism $f : i \to j$ of $\mathcal{J}$ we should have a natural isomorphism between functors $\tau_{i,j}(f) : \pi_i \Rightarrow \pi_j \circ a_f$. In other words, $\tau_{i,j}(f)$ should be a 2-cell in the 2-category $\mathcal{C}$ of small categories. For each $x \in Obj\ A_i$ define $\tau_{i,j}(f)_x : \pi_i(x) = x \to a_f(x) = \pi_j \circ a_f(x)$ to be the isomorphism $h_{(x,f)}$.

LEMMA 4.3. *The map $\pi : F \Rightarrow \Delta_W$ is a pseudo natural transformation with coherence 2-cells given by the natural isomorphisms $\tau$.*

*Proof:* First we show for fixed $f : i \to j$ that the assignment $Obj\ A_i \ni x \mapsto \tau_{i,j}(f)_x \in Mor_W(\pi_i(x), \pi_j \circ a_f(x))$ is a natural transformation. To this end, let $m : x \to y$ be a morphism in the small category $A_i$. By definition, $\tau_{i,j}(f)_x = h_{(x,f)}$, $\tau_{i,j}(f)_y = h_{(y,f)}$, $\pi_i(m) = m$, $\pi_i(x) = x$, $\pi_j \circ a_f(x) = a_f(x)$, and $\pi_j \circ a_f(m) = a_f(m)$. Some similar statements hold for the object $y$. The third requirement on the congruence in $W'$ gives us the following commutative diagram in the small

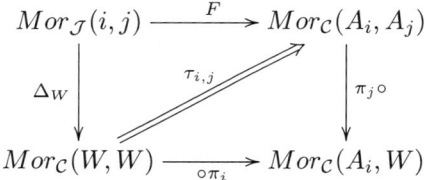

## 4. WEIGHTED PSEUDO COLIMITS IN THE 2-CATEGORY OF SMALL CATEGORIES

category $W$.

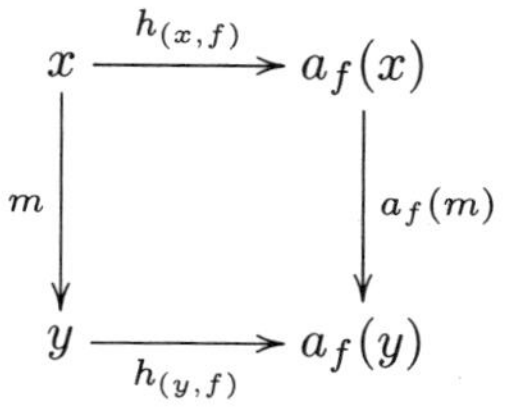

Using the identities just mentioned, the commutivity of this diagram says precisely that $x \mapsto \tau_{i,j}(f)_x$ is a natural transformation. Thus $\tau_{i,j}(f) : \pi_i \Rightarrow \pi_j \circ a_f$ is a natural transformation between functors, i.e. a 2-cell in the 2-category $\mathcal{C}$ of small categories.

The assignment $f \mapsto \tau_{i,j}(f)$ for fixed $i,j$ is natural because the category $Mor_{\mathcal{J}}(i,j)$ has no nontrivial morphisms. Thus $\tau_{i,j}$ is a natural transformation between the indicated functors.

Next we verify the composition axiom for pseudo natural transformations which involves $\tau$ and $\gamma$. The diagram states that $\tau$ must satisfy for all $i \xrightarrow{f} j \xrightarrow{g} k$ in $\mathcal{J}$ the coherence axiom $(i_{\pi_k} * \gamma_{f,g}) \odot (\tau_{j,k}(g) * i_{a_f}) \odot (i_{1_W} * \tau_{i,j}(f)) = \tau_{i,k}(g \circ f) \odot (i_{1_W} * i_{\pi_i})$ as natural transformations. This coherence is satisfied because of the second requirement on the relation in $W'$ for each $x \in Obj\, A_i$ which states $\gamma_{f,g}(x) \circ \tau_{j,k}(g)_{a_f(x)} \circ \tau_{i,j}(f)_x = \tau_{i,k}(g \circ f)_x$. Note that $(i_{\pi_k} * \gamma_{f,g})(x) = \pi_k(\gamma_{f,g}(x)) = \gamma_{f,g}(x)$.

Lastly we verify the unit axiom for pseudo natural transformations which involves $\tau$ and $\delta$. This coherence requires the commutivity of the following diagram for all $j \in Obj\, \mathcal{J}$.

$$\begin{array}{ccccc}
\pi_j & \xrightarrow{i_{\pi_j}} & 1_W \circ \pi_j & \xrightarrow{\delta_{j*}^{\Delta_W} * i_{\pi_j}} & \Delta_W(1_j) \circ \pi_j \\
{\scriptstyle i_{\pi_j}}\Big\Vert & & & & \Big\Vert {\scriptstyle \tau_{1_j} = \tau_{j,j}(1_j)} \\
\pi_j \circ 1_{Fj} & & \xrightarrow{i_{\pi_j} * \delta_{j*}^F} & & \pi_j \circ F(1_j)
\end{array}$$

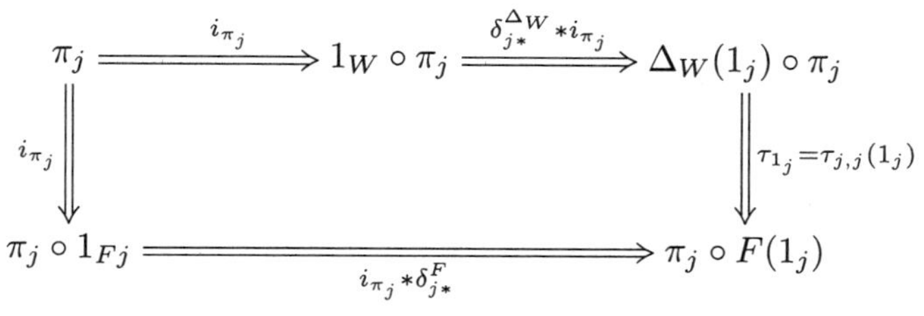

Here $\delta_j^{\Delta_W}$ and $\delta_j^F$ are the natural transformations associated to the pseudo functors $\Delta_W$ and $F$ which make them preserve the identity morphisms $1_j$ up to coherence 2-cell. In fact, $\delta_{j*}^{\Delta_W}$ is trivial. The coherences $\delta_j^{\Delta_W}$ and $\delta_j^F$ fill in the following diagrams for all objects $j$ of $\mathcal{J}$.

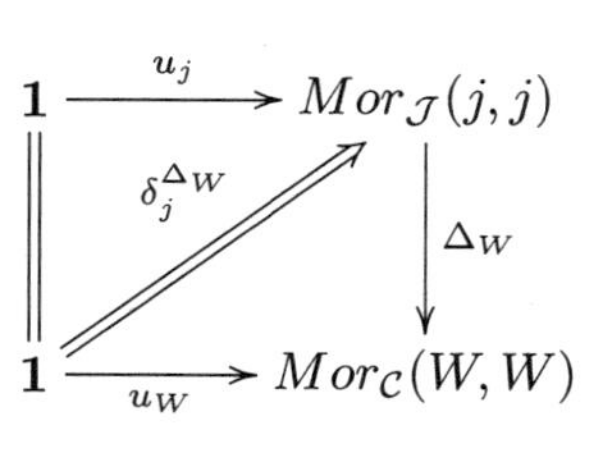

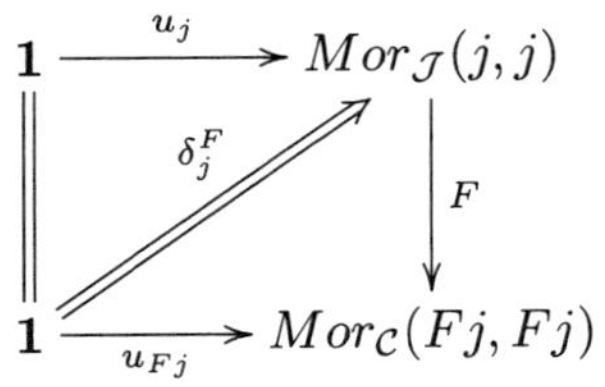


Using the fact that $\delta_j^{\Delta_W}$ evaluated on the unique object $*$ of $\mathbf{1}$ gives the identity 2-cell $i_W : 1_W \Rightarrow 1_W$, the desired coherence diagram simplifies to the following.

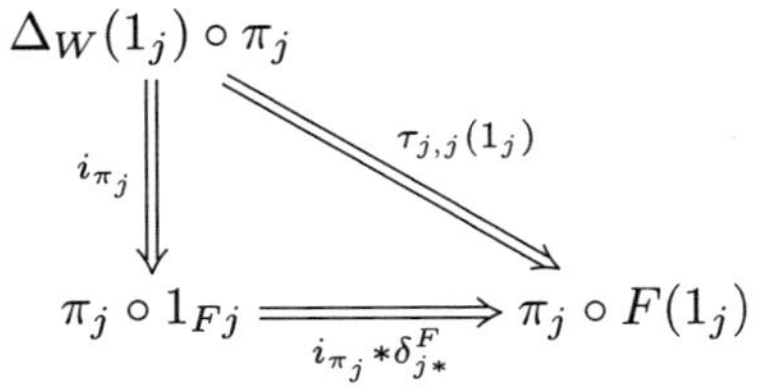


Recall that $(\delta_{j*}^F)_x = h_{(x,1_j)}$ in $W$ by the fourth requirement on the congruence in $W'$. By definition we also have $h_{(x,1_j)} = \tau_{j,j}(1_j)_x$. This implies $(\delta_{j*}^F)_x = h_{(x,1_j)} = \tau_{j,j}(1_j)_x$ and the simplified diagram commutes because $\pi_j$ is the inclusion functor. Hence the required coherence diagram involving $\tau$ and $\delta$ is actually satisfied.

Thus $\pi : F \Rightarrow \Delta_W$ is a pseudo natural transformation with the indicated coherence 2-cells. □

Now we must show that the small category $W$ and the pseudo natural transformation $\pi : F \Rightarrow \Delta_W$ are universal in the sense that the functor $\phi : Mor_{\mathcal{C}}(W,V) \to PseudoCone(F,V)$ defined by $\phi(b) = b \circ \pi$ for objects $b$ is an isomorphism of categories for all objects $V$ of $\mathcal{C}$. More precisely, $\phi$ is defined for $b \in Obj\ Mor_{\mathcal{C}}(W,V)$ and $j \in Obj\ \mathcal{J}$ as $\phi(b)(j) = b \circ \pi_j$. The coherence 2-cells for the pseudo cone $\phi(b)$ are $i_b * \tau_{i,j}(f)$ for all $f : i \to j$ in $\mathcal{J}$. For morphisms $\gamma : b \Rightarrow b'$ in $Mor\ Mor_{\mathcal{C}}(W,V)$ we define $\phi(\gamma) : b \circ \pi \rightsquigarrow b' \circ \pi$ to be the modification which takes $j \in Obj\ \mathcal{J}$ to $\phi(\gamma)(j) = \gamma * i_{\pi_j}$. In the following, $V$ is a fixed object of the 2-category $\mathcal{C}$ of small categories.

LEMMA 4.4. *The map $\phi : Mor_{\mathcal{C}}(W,V) \to PseudoCone(F,V)$ is a functor.*

*Proof:* Let $b \in Obj\ Mor_{\mathcal{C}}(W,V)$ be a functor and $i_b : b \Rightarrow b$ its identity natural transformation. Then obviously $\phi(i_b)(j) = i_b * i_{\pi_j} : b \circ \pi_j \Rightarrow b \circ \pi_j$ is the identity natural transformation $i_{b \circ \pi_j}$ for all $j \in Obj\ \mathcal{J}$ and thus $\phi(i_b)$ is the identity modification. Hence $\phi$ preserves identities.

To verify that $\phi$ preserves compositions, let $\gamma : b \Rightarrow b'$ and $\gamma' : b' \Rightarrow b''$ be natural transformations. Then for each $j \in Obj\ \mathcal{J}$ we have $\phi(\gamma' \odot \gamma)(j) = (\gamma' \odot \gamma) * i_{\pi_j} = (\gamma' \odot \gamma) * (i_{\pi_j} \odot i_{\pi_j})$. By the interchange law we have $(\gamma' \odot \gamma) * (i_{\pi_j} \odot i_{\pi_j}) = (\gamma' * i_{\pi_j}) \odot (\gamma * i_{\pi_j}) = (\phi(\gamma')(j)) \odot (\phi(\gamma)(j)) = (\phi(\gamma') \diamond \phi(\gamma))_j$ where the last equality follows from the definition of vertical composition of modifications. Thus $\phi(\gamma' \odot \gamma) = \phi(\gamma') \diamond \phi(\gamma)$ and $\phi$ preserves compositions. Thus $\phi$ is a functor.

□

The purpose of the next few lemmas is to exhibit an inverse functor $\psi$ for $\phi$.

LEMMA 4.5. *There is a functor $\psi : PseudoCone(F,V) \to Mor_{\mathcal{C}}(W,V)$.*

*Proof:* First we define $\psi$ for objects. Then we define $\psi$ for morphisms. Finally we verify that $\psi$ is a functor.

Let $\pi'$ be an object of $PseudoCone(F,V)$, *i.e.* $\pi' : F \Rightarrow \Delta_V$ is a pseudo natural transformation with coherence 2-cells $\tau'$ up to which $\pi'$ is natural. To define a functor $\psi\pi' = b \in Obj\ Mor_{\mathcal{C}}(W,V)$ we use the universal mapping property of the quotient category $W$ as follows. Define an auxiliary functor $d : W' \to V$ as the functor induced by the map of directed graphs below which is also called $d$.

- For all $i \in Obj\ \mathcal{J}$ and $x \in Obj\ A_i \subseteq Obj\ W'$ let
$$dx := \pi_i' x.$$
- For all $i \in Obj\ \mathcal{J}$, $x, y \in Obj\ A_i$, and all $g \in Mor_{A_i}(x,y) \subseteq Mor_{W'}(x,y)$ let
$$dg := \pi_i' g.$$
- For all $i, j \in Obj\ \mathcal{J}$, $f \in Mor_{\mathcal{J}}(i,j)$, and all $x \in Obj\ A_i \subseteq Obj\ W'$ define
$$d(h_{(x,f)}) := \tau_{i,j}'(f)_x : \pi_i' x \to \pi_j' \circ a_f x$$
$$d(h_{(x,f)}^{-1}) := \tau_{i,j}'(f)_x^{-1} : \pi_j' \circ a_f x \to \pi_i' x.$$

We claim that $d$ preserves the congruence placed on the category $W'$. Following the order in the definition of $\sim$ we have the verifications:

- For $m, n \in Mor\ A_i \subseteq Mor\ W'$ with $Sn = Tm$ we have $d(n \circ_{W'} m) = dn \circ_V dm = \pi_i n \circ_V \pi_i m = \pi_i(n \circ_{A_i} m) = d(n \circ_{A_i} m)$ and for all $1_x \in A_i$ we have $d1_x = \pi_i'(1_x) = 1_{\pi_i' x}$ because $\pi_i'$ is a functor. But $1_{\pi_i' x}$ is also the same as $d$ applied to the identity on $x$ in the free category $W'$.
- Since $\pi'$ is a pseudo natural transformation, for all $i \xrightarrow{f} j \xrightarrow{g} k$ in $\mathcal{J}$ we have
$(i_{\pi_k'} * \gamma_{f,g}) \odot (\tau_{j,k}'(g) * i_{a_f}) \odot (i_{1_V} * \tau_{i,j}'(f)) = \tau_{i,k}'(g \circ f) \odot (i_{1_V} * i_{\pi_i})$ as natural transformations. Evaluating this at $x \in Obj\ A_i$ yields
$$(\pi_k' \gamma_{f,g}(x)) \circ \tau_{j,k}'(g)_{a_f x} \circ \tau_{i,j}'(f)_x = \tau_{i,k}'(g \circ f)_x.$$
This says precisely $d(\gamma_{f,g}(x) \circ_{W'} h_{(a_f(x),g)} \circ_{W'} h_{(x,f)}) = d(h_{(x,g\circ f)})$.
- For all $i, j \in Obj\ \mathcal{J}$, all $f \in Mor_{\mathcal{J}}(i,j)$, and all morphisms $m : x \to y$ of $A_i$ we have to show $d(h_{(y,f)} \circ_{W'} m) = d(a_f(m) \circ_{W'} h_{(x,f)})$. Writing out $d$, we see that this is the same as verifying $\tau_{i,j}'(f)_y \circ_V \pi_i' m = (\pi_j' \circ a_f) m \circ_V \tau_{i,j}'(f)_x$, which is true because the assignment $x \mapsto \tau_{i,j}'(f)_x$ is a natural transformation from $\pi_i'$ to $\pi_j' \circ a_f$.
- For all $j \in Obj\ \mathcal{J}$ and all $x \in Obj\ A_j$ we have to show $d(\delta_{j*}^F)_x = dh_{(x,1_j)}$. Writing out $d$ we see that this is the same as verifying $\pi_j'(\delta_{j*}^F)_x = \tau_{j,j}'(1_j)_x$. Since $\pi'$ is a pseudo natural transformation from $F$ to $\Delta_V$, the natural transformation $\tau'$ must satisfy the coherence $(i_{\pi_j'} * \delta_{j*}^F) \odot i_{\pi_j'} = \tau_{j,j}'(1_j) \odot (i_{1_V} * i_{\pi_j'}) \odot i_{\pi_j'}$ as natural transformations. Evaluating this coherence at $x \in Obj\ A_j$ we get $\pi_j'(\delta_{j*}^F)_x \circ 1_{\pi_j' x} = \tau_{j,j}'(1_j)_x \circ 1_{\pi_j' x} \circ 1_{\pi_j' x}$, which implies $d(\delta_{j*}^F)_x = dh_{(x,1_j)}$ by the remarks above.
- For all $i, j \in Obj\ \mathcal{J}$, $f \in Mor_{\mathcal{J}}(i,j)$, and all $x \in Obj\ A_i \subseteq Obj\ W'$ we have $d(h_{(x,f)}^{-1} \circ_{W'} h_{(x,f)}) = \tau_{i,j}'(f)_x^{-1} \circ \tau_{i,j}'(f)_x = 1_{\pi_j' x} = d(1_x)$ and similarly $d(h_{(x,f)} \circ_{W'} h_{(x,f)}^{-1}) = d(1_{a_f x})$.

Thus $d : W' \to V$ is a functor that preserves the congruence on $W'$. By the universal mapping property of quotient category $W$ of $W'$, there exists a unique functor $b : W \to V$ which factors $d$ via the projection. Define $\psi(\pi') := b \in Obj\ Mor_{\mathcal{C}}(W, V)$. This is how $\psi$ is defined on the objects of the category $PseudoCone(F, V)$.

Next we define $\psi$ on morphisms of the category $PseudoCone(F, V)$. Let $\Xi : \sigma \rightsquigarrow \sigma'$ be a morphism in $PseudoCone(F, V)$, *i.e.* $\Xi$ is a modification from the pseudo natural transformation $\sigma : F \Rightarrow \Delta_V$ to the pseudo natural transformation $\sigma' : F \Rightarrow \Delta_V$. Let $\tau$ and $\tau'$ respectively denote the natural transformations that make the pseudo natural transformations $\sigma$ and $\sigma'$ natural up to cell. We define a morphism $\psi(\Xi)$ of $Mor_{\mathcal{C}}(W, V)$ as follows. Note that such a morphism is by definition a natural transformation between functors from the small category $W$ to the small category $V$. Since $\Xi$ is a modification, we have a 2-cell $\Xi_i : \sigma_i \Rightarrow \sigma_i'$ in the category $\mathcal{C}$ for each $i \in Obj\ \mathcal{J}$. Let $b, b'$ denote the respective functors $\psi(\sigma), \psi(\sigma') : W \to V$. For $x \in Obj\ A_i \subseteq Obj\ W$ define $\psi(\Xi)_x : bx = \sigma_i x \to \sigma_i' x = b'x$ to be $\Xi_i(x) : \sigma_i x \to \sigma_i' x$. The following two commutative diagrams show that $\psi(\Xi)$ is a natural transformation. For $x, y \in Obj\ A_i$ and $m \in Mor_{A_i}(x, y) \subseteq Mor_W(x, y)$ the diagram

$$\begin{array}{ccc} bx & \xrightarrow{\Xi_i x} & b'x \\ {\scriptstyle \sigma_i m = bm}\downarrow & & \downarrow{\scriptstyle b'm = \sigma_i' m} \\ by & \xrightarrow[\Xi_i y]{} & b'y \end{array}$$

in $V$ commutes because $\Xi_i : \sigma_i \Rightarrow \sigma_i'$ is a natural transformation. For a morphism $f : i \to j$ in $\mathcal{J}$ the diagram

$$\begin{array}{ccc} bx & \xrightarrow{\Xi_i x} & b'x \\ {\scriptstyle \tau_{i,j}(f)_x = bh_{(x,f)}}\downarrow & & \downarrow{\scriptstyle b'h_{(x,f)} = \tau'_{i,j}(f)_x} \\ ba_f(x) & \xrightarrow[\Xi_j a_f(x)]{} & b'a_f(x) \end{array}$$

commutes because of the coherence in the definition of modification and because of the definitions of $b, b'$ on $h_{(x,f)}$. We see this by taking $\gamma = i_f$ in diagrams (3.1) and (3.2) in the definition of modification. An inductive argument shows that $\psi(\Xi)$ is natural for all other arrows in $W$ as well. Hence $\psi(\Xi) : \psi(\sigma) \Rightarrow \psi(\sigma')$ is a morphism in the category $Mor_{\mathcal{C}}(W, V)$.

Lastly we verify that $\psi$ is a functor, *i.e.* that $\psi$ preserves the identity modifications and the composition of modifications. Let $\Xi : \sigma \rightsquigarrow \sigma$ be the identity modification belonging to a pseudo natural transformation $\sigma : F \Rightarrow \Delta_V$. This means that $\Xi_i : \sigma_i \Rightarrow \sigma_i$ is the identity natural transformation for the functor $\sigma_i : A_i \to V$. For all $i \in Obj\ \mathcal{J}$ and all $x \in Obj\ A_i$ we have by definition of $\psi$ that $\psi(\Xi)_x : \psi(\sigma)x = \sigma_i x \to \sigma_i x = \psi(\sigma)x$ is $\Xi_i(x) : \sigma_i x \to \sigma_i x$, which is the identity morphism on the object $\sigma_i x$ of the small category $V$ by hypothesis. Hence $\psi(\Xi) : \psi(\sigma) \to \psi(\sigma)$ is the identity natural transformation and $\psi$ preserves identity modifications.

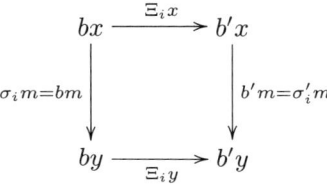

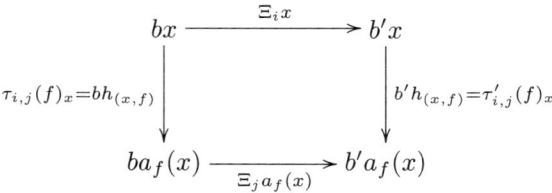

To verify that $\psi$ preserves compositions, let $\Xi : \sigma \rightsquigarrow \sigma'$ and $\Xi' : \sigma' \rightsquigarrow \sigma''$ be modifications. Then the vertical composition of modifications (which makes $PseudoCone(F, V)$ a category) is defined as $(\Xi' \diamond \Xi)_i := \Xi'_i \odot \Xi_i$ where $\Xi'_i \odot \Xi_i$ is the vertical composition of the natural transformations $\Xi_i : \sigma_i \Rightarrow \sigma'_i$ and $\Xi'_i : \sigma'_i \Rightarrow \sigma''_i$ as usual. Then for all $i \in Obj\ \mathcal{J}$ and all $x \in Obj\ A_i \subseteq Obj\ W$ we have $\psi(\Xi' \diamond \Xi)_x = (\Xi' \diamond \Xi)_i(x) = (\Xi'_i \odot \Xi_i)_x = \Xi'_i(x) \circ \Xi_i(x) = \psi(\Xi')_x \circ \psi(\Xi)_x = (\psi(\Xi') \odot \psi(\Xi))_x$. Thus $\psi(\Xi' \diamond \Xi) = \psi(\Xi') \odot \psi(\Xi)$ and $\psi$ preserves compositions of modifications. Hence $\psi$ is a functor. $\square$

LEMMA 4.6. *The functor $\phi \circ \psi : PseudoCone(F, V) \to PseudoCone(F, V)$ is the identity functor.*

*Proof:* First we verify this for objects, then for morphisms. Let $\pi' : F \Rightarrow \Delta_V$ be a pseudo natural transformation with coherence isomorphisms $\tau'$. Let $b = \psi(\pi')$. Then using the definitions of $b$ in Lemma 4.5 and the definition of $\pi$ above we evaluate $\phi(\psi(\pi'))$ at each object $i$ of $\mathcal{J}$ and compare the resulting functor $\phi(\psi(\pi'))_i$ to the functor $\pi'_i$. Formally this is:

- For all $x \in Obj\ A_i$, we have
$$\phi(\psi(\pi'))_i x = \phi(b)_i x = (b \circ \pi_i)x = bx = \pi'_i x.$$
- For all $x, y \in Obj\ A_i$ and all $g \in Mor_{A_i}(x, y)$ we have
$$\phi(\psi(\pi'))_i g = \phi(b)_i g = (b \circ \pi_i)g = bg = \pi'_i g.$$

Thus $\phi(\psi(\pi')) = \pi'$ for all objects $\pi'$ of the category $PseudoCone(F, V)$. Hence $\phi \circ \psi$ is the identity on objects.

Next we verify the lemma for morphisms. Let $\Xi : \sigma \rightsquigarrow \sigma'$ be a morphism in the category $PseudoCone(F, V)$, *i.e.* $\Xi$ is a modification from the pseudo natural transformation $\sigma : F \Rightarrow \Delta_V$ to the pseudo natural transformation $\sigma' : F \Rightarrow \Delta_V$. Let $b = \psi(\sigma), b' = \psi(\sigma') : W \to V$ and $\gamma = \psi(\Xi) : b \Rightarrow b'$ for more convenient notation. Then $\phi(\psi(\Xi)) = \phi(\gamma) : b \circ \pi \rightsquigarrow b' \circ \pi$ is a modification from $\sigma$ to $\sigma'$ by the result on objects. For each $j \in Obj\ \mathcal{J}$ we have the natural transformation $\phi(\gamma)(j) = \gamma * i_{\pi_j} : b \circ \pi_j \Rightarrow b' \circ \pi_j$. But this natural transformation is precisely $\Xi_j : \sigma_j \Rightarrow \sigma'_j$ by the definition of $\gamma$ via $\psi$. Thus for all morphisms $\Xi$ of the category $PseudoCone(F, V)$ we have $\phi(\psi(\Xi)) = \Xi$. Hence $\phi \circ \psi$ is the identity on morphisms. $\square$

LEMMA 4.7. *The composite functor $\psi \circ \phi : Mor_{\mathcal{C}}(W, V) \to Mor_{\mathcal{C}}(W, V)$ is the identity functor.*

*Proof:* First we verify this for objects, then on generators for morphisms. Let $b : W \to V$ be a functor and $x \in Obj\ A_i \subseteq Obj\ W$. Then $\psi \circ \phi(b)x = \psi(b \circ \pi)x = (b \circ \pi_i)x = bx$. Similarly for a morphism $g \in Mor_{A_i}(x, y) \subseteq Mor_W(x, y)$ we have $\psi \circ \phi(b)g = \psi(b \circ \pi)g = (b \circ \pi_i)g = bg$. For morphisms $h_{(x,f)}$, the analogous calculation is $\psi \circ \phi(b)h_{(x,f)} = \psi(b \circ \pi)h_{(x,f)} = (i_b * \tau_{i,j}(f))_x = b(\tau_{i,j}(f)_x) = bh_{(x,f)}$. That follows because the coherence 2-cell up to which $b \circ \pi$ is natural is $(i_b * \tau_{i,j}(f))_x = b(\tau_{i,j}(f)_x)$, then we use the third part of the definition of $\psi$ as well as the definition $h_{(x,f)} = \tau_{i,j}(f)_x$. Thus $\psi \circ \phi(b) = b$ for all objects $b$ of the category $Mor_{\mathcal{C}}(W, V)$. Hence $\psi \circ \phi$ is the identity on the objects of the category $Mor_{\mathcal{C}}(W, V)$.

Next we verify the lemma for morphisms. Let $\gamma : b \Rightarrow b'$ be a morphism in $Mor_{\mathcal{C}}(W, V)$, *i.e.* a natural transformation from some functor $b$ to some functor

$b'$. Let $\Xi = \phi(\gamma)$, $\sigma = \phi(b)$, and $\sigma' = \phi(b')$ for more convenient notation. Then by definition $\Xi : \sigma = b \circ \pi \rightsquigarrow b' \circ \pi = \sigma'$ is the modification which takes $j \in \mathcal{J}$ to $\gamma * i_{\pi_j}$. Let $x \in Obj\ A_j \subseteq Obj\ W$. Then $\psi(\Xi)_x : \psi(\sigma)x = \sigma_j x \to \sigma'_j x = \psi(\sigma')x$ is $\Xi_j(x) = (\gamma * i_{\pi_j})_x : (b \circ \pi)_j x \to (b' \circ \pi)_j x$. This is described by the following diagram.

$$\begin{array}{ccccc} A_j & \xrightarrow{\pi_j} & W & \xrightarrow{b} & V \\ & & \Big\Downarrow i_{\pi_j} & & \Big\Downarrow \gamma \\ A_j & \xrightarrow[\pi_j]{} & W & \xrightarrow[b']{} & V \end{array}$$

But by definition of $\phi$ and $(b \circ \pi)_j$, we see that $\Xi_j(x) = \gamma_{\pi_j x} = \gamma_x$ is precisely $\gamma_x : bx \to b'x$. Thus $\psi(\Xi)_x = \gamma_x$ and $\psi(\phi(\gamma)) = \psi(\Xi) = \gamma$. Hence $\psi \circ \phi$ is the identity on the morphisms of the category $Mor_{\mathcal{C}}(W, V)$. □

LEMMA 4.8. *The small category $W$ and the pseudo natural transformation $\pi : F \Rightarrow \Delta_W$ are universal in the sense that the functor $\phi : Mor_{\mathcal{C}}(W, V) \to PseudoCone(F, V)$ defined by $\phi b = b \circ \pi$ for objects $b$ is an isomorphism of categories for all objects $V$ of $\mathcal{C}$.*

*Proof:* This follows immediately from the previous four lemmas because $V$ was an arbitrary object of the 2-category $\mathcal{C}$. □

LEMMA 4.9. *The small category $W$ and the pseudo natural transformation $\pi : F \Rightarrow \Delta_W$ are a pseudo colimit of the pseudo functor $F : \mathcal{J} \to \mathcal{C}$.*

*Proof:* This follows from Lemmas 4.3 and 4.8. □

Thus every pseudo functor $F : \mathcal{J} \to \mathcal{C}$ from a small 1-category $\mathcal{J}$ to the 2-category $\mathcal{C}$ of small categories admits a pseudo colimit. In other words, the 2-category $\mathcal{C}$ of small categories admits pseudo colimits. This completes the proof of Theorem 4.2. □

LEMMA 4.10. *The 2-category of $\mathcal{C}$ of small categories admits tensor products.*

*Proof:* Let $J$ and $F$ be small categories. Then $J * F := J \times F$ is a tensor product of $J$ and $F$ with unit $\pi : J \to Cat(F, J \times F)$ defined by

$$\pi(j)(x) := (j, x)$$

$$\pi(j)(f) := (1_j, f)$$

$$\pi(g)_x := (g, 1_x)$$

for $j \in Obj\ J, x \in Obj\ F, f \in Mor\ F, g \in Mor\ J$. Alternatively one can see that $Cat(J \times F, C)$ is isomorphic to $Cat(J, Cat(F, C))$ by the usual adjunction. □

LEMMA 4.11. *The 2-category $\mathcal{C}$ of small categories admits weighted pseudo colimits.*

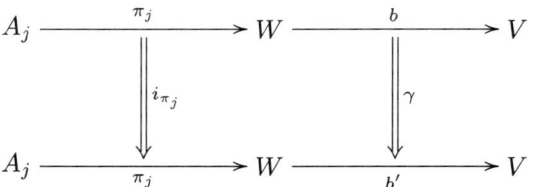

*Proof:* This 2-category admits pseudo coequalizers by Theorem 4.2. It also admits tensor products by Lemma 4.10. It is not difficult to construct 2-coproducts in this 2-category by using disjoint union. Hence, by the dual version of Theorem 3.22, the 2-category $\mathcal{C}$ admits weighted pseudo limits. □

REMARK 4.12. The 2-category of small groupoids admits weighted pseudo colimits.

*Proof:* The proof is the same as in the proof for the 2-category of small categories except that we replace the free category by the free groupoid. □

THEOREM 4.13. *The 2-category of small categories and the 2-category of small groupoids admit weighted bicolimits.*

*Proof:* These 2-categories admit weighted pseudo colimits. Every weighted pseudo colimit is a weighted bicolimit. □

CHAPTER 5

# Weighted Pseudo Limits in the 2-Category of Small Categories

Not only does the 2-category $\mathcal{C}$ of small categories admit pseudo colimits, but it also admits pseudo limits. In fact we construct them explicitly in the next proof. The notation remains the same as in the previous chapter. This description is not new, since the candidate $L$ in the proof below can be found in [**50**]. Theorem 3.22 allows us to conclude that $\mathcal{C}$ admits weighted pseudo limits.

THEOREM 5.1. *The 2-category $\mathcal{C}$ of small categories admits pseudo limits.*

*Proof:* Let $\mathcal{J}$ be a small 1-category and $F : \mathcal{J} \to \mathcal{C}$ a pseudo functor. Recall that a pseudo cone from $X$ to $F$ is a pseudo natural transformation $\Delta_X \Rightarrow F$ and that $PseudoCone(X, F)$ denotes the category with objects the pseudo cones from $X$ to $F$ and morphisms the modifications between them. A pseudo limit of $F$ is an object $L \in Obj\ \mathcal{C}$ with a pseudo cone $\pi : \Delta_L \Rightarrow F$ which are universal in the sense that $(\pi\circ) : Mor_{\mathcal{C}}(V, L) \to PseudoCone(V, F)$ is an isomorphism of categories for all small categories $V$.

First we define candidates $L \in Obj\ \mathcal{C}$ and $\pi : \Delta_L \Rightarrow F$. Then we show that they are universal. For each $j \in Obj\ \mathcal{J}$ let $A_j$ denote the small category $Fj$ as in the proof for the pseudo colimit. Then the candidate for the pseudo limit is $L := PseudoCone(\mathbf{1}, F)$, also called the category of pseudo cones to $F$ on a point. The pseudo natural transformation candidate $\pi : \Delta_L \Rightarrow F$ is defined for all objects $\eta : \Delta_{\mathbf{1}} \Rightarrow F$ of $L$ as $\pi_i(\eta) := \eta_i(*)$ for all $i \in Obj\ \mathcal{J}$. For morphisms $\Theta : \eta \rightsquigarrow \eta'$ of $L$ define $\pi_i(\Theta) := \Theta_i(*) : \eta_i(*) \to \eta'_i(*)$ for all $i \in Obj\ \mathcal{J}$. Define the coherence isos $\tau_{i,j}$

$$\begin{array}{ccc} Mor_{\mathcal{J}}(i,j) & \xrightarrow{\Delta_L} & Mor_{\mathcal{C}}(L,L) \\ {\scriptstyle F}\downarrow & \overset{\tau_{i,j}}{\nearrow\!\!\!\nearrow} & \downarrow{\scriptstyle \pi_j\circ} \\ Mor_{\mathcal{C}}(A_i,A_j) & \xrightarrow[\circ\pi_i]{} & Mor_{\mathcal{C}}(L,A_j) \end{array}$$

belonging to $\pi : \Delta_L \Rightarrow F$ by $\tau_{i,j}(f)_\eta := \tau^\eta_{i,j}(f)_*$ for all $f \in Mor_{\mathcal{J}}(i,j)$ and all $\eta \in Obj\ L$ where $\tau^\eta_{i,j}$ is the coherence natural isomorphism belonging to $\eta : \Delta_{\mathbf{1}} \Rightarrow F$.

$$\begin{array}{ccc} Mor_{\mathcal{J}}(i,j) & \xrightarrow{\Delta_{\mathbf{1}}} & Mor_{\mathcal{C}}(\mathbf{1},\mathbf{1}) \\ {\scriptstyle F}\downarrow & \overset{\tau^\eta_{i,j}}{\nearrow\!\!\!\nearrow} & \downarrow{\scriptstyle \eta_j\circ} \\ Mor_{\mathcal{C}}(A_i,A_j) & \xrightarrow[\circ\eta_i]{} & Mor_{\mathcal{C}}(\mathbf{1},A_j) \end{array}$$

LEMMA 5.2. *The map $\pi : \Delta_L \Rightarrow F$ is a pseudo natural transformation with coherence 2-cells given by $\tau$.*

*Proof:* First we show that for each $j \in Obj\ \mathcal{J}$ we have a morphism $\pi_j : L = \Delta_L(j) \to Fj = A_j$ in the 2-category $\mathcal{C}$. We claim that $\pi_j$ is a morphism, *i.e.* a functor. Let $1_\eta = \Theta : \eta \rightsquigarrow \eta$ be the identity modification of the pseudo cone $\eta : \Delta_{\mathbf{1}} \Rightarrow F$. This means $\Theta_j = i_{\eta_j} : \eta_j \Rightarrow \eta_j$ is the identity natural transformation for all $j \in Obj\ \mathcal{J}$. Then $\pi_j(1_\eta) = \pi_j(\Theta) = \Theta_j(*) = 1_{\eta_j(*)} = 1_{\pi_j(\eta)}$ and $\pi_j$ preserves identities. Now let $\Theta, \Xi$ denote modifications in $L$ such that $\Xi \diamond \Theta$ exists. Then $\pi_j(\Xi \diamond \Theta) = (\Xi \diamond \Theta)_j(*) = \Xi_j \odot \Theta_j(*) = \Xi_j(*) \circ \Theta_j(*) = \pi_j(\Xi) \circ \pi_j(\Theta)$. Thus $\pi_j : L \to A_j$ is a functor.

Next we show that $\tau_{i,j}$ as defined above is a natural transformation for all $i, j \in Obj\ \mathcal{J}$. By inspecting the definition diagram for $\tau_{i,j}$ above we see that for all $f \in Mor_{\mathcal{J}}(i,j)$ we should have an element $\tau_{i,j}(f)$ of $Mor\ Mor_{\mathcal{C}}(L, A_j)$. To this end, we claim that $\tau_{i,j}(f) : Ff \circ \pi_i \Rightarrow \pi_j$ is a natural transformation. To see this, let $\Theta : \eta \rightsquigarrow \eta'$ be a modification, *i.e.* a morphism in the category $L$. Then by taking $\gamma = i_f$ in the definition of modification and evaluating the modification diagrams (3.1) and (3.2) at $* \in Obj\ \mathbf{1}$ with $\alpha = \eta, \beta = \eta', A = i, B = j, \Xi = \Theta$ we obtain the commutivity of the diagram in the category $A_j$

$$\begin{array}{ccc}
Ff(\eta_i(*)) & \xrightarrow{\tau^{\eta}_{i,j}(f)_*} & \eta_j(*) \\
{\scriptstyle Ff(\Theta_i(*))}\downarrow & & \downarrow{\scriptstyle \Theta_j(*)} \\
Ff(\eta'_i(*)) & \xrightarrow[\tau^{\eta'}_{i,j}(f)_*]{} & \eta'_j(*)
\end{array}$$

where $\tau^\eta$ and $\tau^{\eta'}$ denote the coherence natural transformations belonging to the pseudo cones $\eta$ and $\eta'$ respectively. Using the definitions $\tau_{i,j}(f)_\eta := \tau^{\eta}_{i,j}(f)_*, \pi_i(\eta) := \eta_i(*)$, and $\pi_i(\Theta) := \Theta_i(*)$ we see that this diagram is

$$\begin{array}{ccc}
Ff \circ \pi_i(\eta) & \xrightarrow{\tau_{i,j}(f)_\eta} & \pi_j(\eta) \\
{\scriptstyle Ff\circ\pi_i(\Theta)}\downarrow & & \downarrow{\scriptstyle \pi_j(\Theta)} \\
Ff \circ \pi_i(\eta') & \xrightarrow[\tau_{i,j}(f)_{\eta'}]{} & \pi_j(\eta')
\end{array}$$

which says precisely that $\eta \mapsto \tau_{i,j}(f)_\eta$ is natural for fixed morphisms $f : i \to j$ of $\mathcal{J}$. Thus $\tau_{i,j}(f) : Ff \circ \pi_i \Rightarrow \pi_j$ is a natural transformation. On the other hand, the assignment $Mor_{\mathcal{J}}(i,j) \ni f \mapsto \tau_{i,j}(f)$ is vacuously natural because the category $Mor_{\mathcal{J}}(i,j)$ is discrete. Thus $\tau_{i,j}$ is a natural transformation for all $i, j \in Obj\ \mathcal{J}$.

The natural isomorphisms $\tau$ satisfy the unit axiom and composition axiom involving $\delta$ and $\gamma$ because the individual $\tau^\eta$ do. □

Now we must show that the small category $L$ and the pseudo natural transformation $\pi : \Delta_L \Rightarrow F$ are universal in the sense that the functor $\phi : Mor_{\mathcal{C}}(V, L) \to PseudoCone(V, F)$ defined by $\phi b = \pi \circ b$ for objects $b$ is an isomorphism of categories for all objects $V$ of $\mathcal{C}$. More precisely, $\phi$ is defined for $b \in Obj\ Mor_{\mathcal{C}}(V, L)$ and $j \in Obj\ \mathcal{J}$ as $\phi(b)(j) = \pi_j \circ b$. The natural transformations for the pseudo cone $\phi b$

are $\tau_{i,j}(f) * i_b$ for all $f : i \to j$ in $\mathcal{J}$. For morphisms $\gamma : b \Rightarrow b'$ in $Mor\ Mor_{\mathcal{C}}(V, L)$ we define $\phi(\gamma) : \pi \circ b \rightsquigarrow \pi \circ b'$ to be the modification which takes $j \in Obj\ \mathcal{J}$ to $\phi(\gamma)(j) = i_{\pi_j} * \gamma$. In the following, $V$ is a fixed object of the 2-category $\mathcal{C}$ of small categories.

LEMMA 5.3. *The map* $\phi : Mor_{\mathcal{C}}(V, L) \to PseudoCone(V, F)$ *is a functor.*

*Proof:* The proof is analogous to the proof for the $\phi$ of the pseudo colimit. □

Now we construct a functor $\psi : PseudoCone(V, F) \to Mor_{\mathcal{C}}(V, L)$ that is inverse to $\phi$. First we define $\psi$ for objects, then for morphism. Finally we verify that it is a functor and inverse to $\phi$. The key observation in the construction is that we can get a pseudo cone on a point by evaluating a pseudo cone on an object. This is the essence of the identification we make below.

REMARK 5.4. Let $Obj\ P$ be the subset of the set
$\{(a_i)_i \times (\varepsilon_f)_f \in \prod_{i \in Obj\ \mathcal{J}} Obj\ A_i \times \prod_{f \in Mor\ \mathcal{J}} Mor\ A_{Tf} |\ \varepsilon_f : Ff(a_{Sf}) \to a_{Tf}$ is iso
for all $f \in Mor\ \mathcal{J}\}$ consisting of all $(a_i)_i \times (\varepsilon_f)_f$ such that:

- $\varepsilon_{1_j} \circ \delta^F_{j*}(a_j) = 1_{a_j}$ for all $j \in Obj\ \mathcal{J}$.
- $\varepsilon_g \circ (Fg(\varepsilon_f)) = \varepsilon_{g \circ f} \circ \gamma^F_{f,g}(a_{Sf})$ for all $f, g \in Mor\ \mathcal{J}$ such that $g \circ f$ exists.

Then $Obj\ L$ and $Obj\ P$ are in bijective correspondence via the map $Obj\ L \to Obj\ P$, $\eta \mapsto (\eta_i(*))_i \times (\tau^{\eta}_{Sf,Tf}(f)_*)_f$.

*Proof:* The two conditions express exactly the required coherences for a pseudo cone $\eta : \Delta_1 \Rightarrow F$. Any pseudo cone $\eta : \Delta_1 \Rightarrow F$ is completely determined by the data listed in the image sequence. □

REMARK 5.5. Let $\eta = (a_i)_i \times (\varepsilon_f)_f$ and $\eta' = (a'_i)_i \times (\varepsilon'_f)_f$ be elements of $Obj\ P$. Let $Mor_P(\eta, \eta')$ denote the set of $(\xi_i)_i \in \prod_{i \in Obj\ \mathcal{J}} Mor_{A_i}(a_i, a'_i)$ such that

$$\begin{array}{ccc} Ff(a_i) & \xrightarrow{\varepsilon_f} & a_j \\ {\scriptstyle Ff(\xi_i)} \downarrow & & \downarrow {\scriptstyle \xi_j} \\ Ff(a'_i) & \xrightarrow[\varepsilon'_f]{} & a'_j \end{array}$$

commutes for all $f : i \to j$ in $\mathcal{J}$. Then $Mor_L(\eta, \eta')$ and $Mor_P(\eta, \eta')$ are in bijective correspondence via the map $Mor_L(\eta, \eta') \to Mor_P(\eta, \eta')$, $\Theta \mapsto (\Theta_i(*))_i$. Moreover, the composition $\Theta \diamond \Xi$ in $Mor_L(\eta, \eta')$ corresponds to the componentwise composition in $Mor_P(\eta, \eta')$.

*Proof:* The diagram is the result of evaluating the coherence stated in diagrams (3.1) and (3.2) in the definition of modification at $*$. The claim about composition follows immediately from the definition of vertical composition $\diamond$ of modifications. □

REMARK 5.6. Under the identification above, $P$ is a category and $\pi_j$ is the projection onto the $j$-th coordinate.

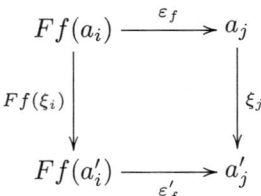

*Proof:* This follows directly from the definition of $\pi$ and the identification. $\square$

We will use the identification without explanation. Now we define a functor $\psi(\pi') = b : V \to L$ for any object $\pi'$ of $PseudoCone(V, F)$. This will substantiate the comment that evaluating a pseudo cone on an object gives a pseudo cone on a point.

LEMMA 5.7. *Let $\pi' : \Delta_V \Rightarrow F$ be a pseudo natural transformation with coherence natural isomorphisms $\tau'$. For any fixed $x \in Obj\ V$ we have $\psi(\pi')(x) := b(x) := (\pi_i'(x))_i \times (\tau'_{Sf,Tf}(f)_x)_f$ is an element of $Obj\ P = Obj\ L$.*

*Proof:* Evaluating the coherences for $\tau$ involving $\delta$ and $\gamma$ at the object $x$ gives the coherences in the definition of $P$. Thus $b(x) \in Obj\ P$ and $b(x)$ is a pseudo cone $\Delta_{\mathbf{1}} \Rightarrow F$, in other words $b(x)$ is a pseudo cone on a point. $\square$

LEMMA 5.8. *Let $\pi' : \Delta_V \Rightarrow F$ be a pseudo natural transformation with coherence natural isomorphisms $\tau'$. Then for any fixed $h \in Mor_V(x, y)$ we have a modification $\psi(\pi')(h) := b(h) := (\pi_i'(h))_i : b(x) \rightsquigarrow b(y)$. This notation means $b(h)_i(*) := \pi_i'(h)$.*

*Proof:* For notational convenience let $\eta := b(x) : \Delta_{\mathbf{1}} \Rightarrow F$ and $\eta' := b(y) : \Delta_{\mathbf{1}} \Rightarrow F$. Let $\Theta = b(h)$. Then $\tau'_{i,j}(f)_x = \tau^{\eta}_{i,j}(f)_*$ and $\tau'_{i,j}(f)_y = \tau^{\eta'}_{i,j}(f)_*$ and $\Theta_i(*) = \pi_i'(h)$ for all $f : i \to j$ in $\mathcal{J}$ by the identification. The naturality of $\tau'_{i,j}(f)$ says $\tau'_{i,j}(f)_y \circ Ff(\pi_i'(h)) = \pi_j'(h) \circ \tau'_{i,j}(f)_x$ for all $f : i \to j$ in $\mathcal{J}$. Rewriting this identity using $\eta, \eta'$, and $\Theta$ gives $\tau^{\eta'}_{i,j}(f)_* \circ Ff(\Theta_i(*)) = \Theta_j(*) \circ \tau^{\eta}_{i,j}(f)_*$. This last identity says that the composition of natural transformations (2-cells)

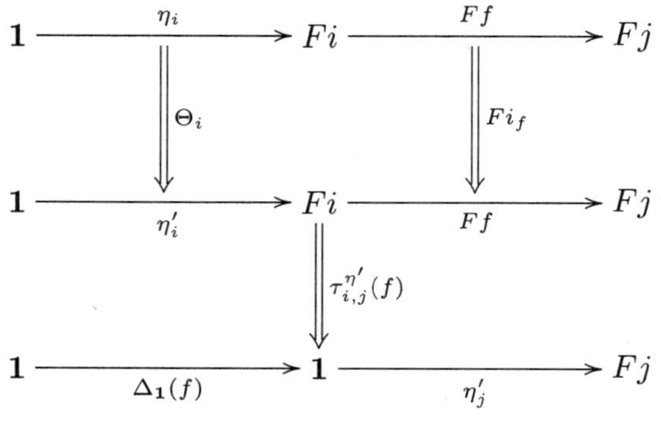

is the same as the composition

$$\begin{array}{ccccc}
\mathbf{1} & \xrightarrow{\eta_i} & Fi & \xrightarrow{Ff} & Fj \\
 & & \Big\Downarrow {\scriptstyle \tau^{\eta}_{i,j}(f)} & & \\
\mathbf{1} & \xrightarrow{\Delta_{\mathbf{1}}(f)} & \mathbf{1} & \xrightarrow{\eta_j} & Fj \\
 & \Big\Downarrow {\scriptstyle \Delta_{\mathbf{1}}(i_f)} & & \Big\Downarrow {\scriptstyle \Theta_j} & \\
\mathbf{1} & \xrightarrow[\Delta_{\mathbf{1}}(f)]{} & \mathbf{1} & \xrightarrow[\eta'_j]{} & Fj
\end{array}$$

of natural transformations for all $f : i \longrightarrow j$ in $\mathcal{J}$. The only 2-cells in the category $\mathcal{J}$ are of the form $i_f$. Therefore we have verified diagrams (3.1) and (3.2) for $\Theta$ to be a modification. Thus $\psi(\pi')(h) = b(h) = \Theta : \eta \rightsquigarrow \eta'$ is a modification. $\square$

LEMMA 5.9. *For any pseudo natural transformation $\pi' : \Delta_V \Rightarrow F$ the map $\psi(\pi') = b : V \longrightarrow L$ is a functor.*

*Proof:* For each $x \in Obj\ V$ and all $j \in Obj\ \mathcal{J}$ we have $b(1_x)_j(*) = \pi'_j(1_x) = 1_{\pi'_j x}$ since $\pi'_j : V \longrightarrow A_j$ is a functor. Hence $b(1_x)_j = i_{b(x)_j}$. Hence $b(1_x) : b(x) \rightsquigarrow b(x)$ is the identity modification. If $h$ and $\ell$ are morphisms in $V$ such that $\ell \circ h$ exists, then $b(\ell \circ h)_j(*) = \pi'_j(\ell \circ h) = \pi'_j(\ell) \circ \pi'_j(h) = b(\ell)_j(*) \circ b(h)_j(*) = (b(\ell)_j \odot b(h)_j)(*) = (b(\ell) \diamond b(h))_j(*)$. Hence $b(\ell \circ h) = b(\ell) \diamond b(h)$ and $b$ preserves compositions. $\square$

LEMMA 5.10. *Let $\Xi : \alpha \rightsquigarrow \beta$ be a morphism in the category $PseudoCone(V, F)$. Then $\psi(\Xi) : \psi(\alpha) \Rightarrow \psi(\beta)$ defined by $V \ni x \mapsto (\Xi_i(x))_i \in Mor_L(\psi(\alpha)x, \psi(\beta)x)$ is a natural transformation. As in Remark 5.5 above, this definition means $\psi(\Xi)(x)_i(*) := \Xi_i(x)$.*

*Proof:* Since $\Xi : \alpha \rightsquigarrow \beta$ is a modification, for each object $i$ of $\mathcal{J}$ there is a 2-cell of $\mathcal{C}$ (a natural transformation) $\Xi_i : \alpha_i \Rightarrow \beta_i$ and these satisfy the condition listed in the definition of modification. Evaluating this condition in diagrams (3.1) and (3.2) at $x \in V$ we see that $(\Xi_i(x))_i : \psi(\alpha)x \rightsquigarrow \psi(\beta)x$ is a modification. Hence $(\Xi_i(x))_i \in Mor_L(\psi(\alpha)x, \psi(\beta)x)$.

We claim that $\psi(\Xi)$ is natural, *i.e.* that the diagram

$$\begin{array}{ccc}
\psi(\alpha)x & \xrightarrow{(\Xi_i(x))_i} & \psi(\beta)x \\
{\scriptstyle \psi(\alpha)g=(\alpha_i(g))_i} \Big\downarrow & & \Big\downarrow {\scriptstyle (\beta_i(g))_i=\psi(\beta)g} \\
\psi(\alpha)y & \xrightarrow[(\Xi_i(y))_i]{} & \psi(\beta)y
\end{array}$$

in $L$ commutes. We only need to verify that the diagram commutes componentwise, since the vertical composition of modifications corresponds to the componentwise composition of these sequences under the identification. But the diagram obviously commutes componentwise because $\Xi_i : \alpha_i \Rightarrow \beta_i$ is a natural transformation. $\square$

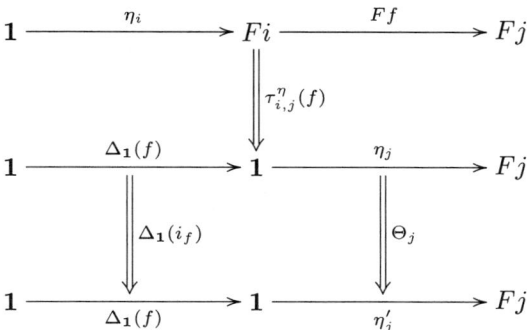

THEOREM 5.11. *The map* $\psi : PseudoCone(V, F) \to Mor_{\mathcal{C}}(V, L)$ *as defined in the previous lemmas is a functor.*

*Proof:* Suppose $\Xi : \alpha \rightsquigarrow \alpha$ is the identity modification for a pseudo cone $\alpha : \Delta_V \Rightarrow F$. Then $\Xi_j = i_{\alpha_j} : \alpha_j \Rightarrow \alpha_j$ for all $j \in Obj\ \mathcal{J}$, so that $\Xi_j(x) = (i_{\alpha_j})_x = 1_{\alpha_j(x)}$. Then $x \mapsto (1_{\alpha_j(x)})_j$ is the identity morphism $\psi(\alpha) \to \psi(\alpha)$ in $Mor_{\mathcal{C}}(V, L)$.

If $\Xi, \Theta$ are modifications in $PseudoCone(V, F)$ such that $\Theta \diamond \Xi$ exists, then for all $x \in V$ we have

$$\begin{aligned}\psi(\Theta \diamond \Xi)(x) &= ((\Theta \diamond \Xi)_i(x))_i \\ &= ((\Theta_i \odot \Xi_i)(x))_i \\ &= (\Theta_i(x) \circ \Xi_i(x))_i \\ &= (\Theta_i(x))_i \diamond (\Xi_i(x))_i \\ &= \psi(\Theta)(x) \diamond \psi(\Xi)(x) \\ &= (\psi(\Theta) \odot \psi(\Xi))(x).\end{aligned}$$

Hence $\psi(\Theta \diamond \Xi) = \psi(\Theta) \odot \psi(\Xi)$ and $\psi$ preserves compositions.

□

Now that we have constructed the functor $\psi$, we prove that it is inverse to $\phi$.

LEMMA 5.12. *The functor* $\psi$ *is a left inverse for* $\phi$, *i.e.* $\psi \circ \phi = 1_{Mor_{\mathcal{C}}(V,L)}$.

*Proof:* First we verify the identity on objects. Let $b : V \to L$ be an object of $Mor_{\mathcal{C}}(V, L)$. Recall that $\phi(b)$ is the pseudo natural transformation $\pi \circ b$ with the coherence natural transformations $\tau'_{i,j}(f) = \tau_{i,j}(f) * i_b$ for all $f : i \to j$. For $x \in V$ we have

$$\begin{aligned}\psi \circ \phi(b)(x) &= \psi(\pi \circ b)(x) \\ &= (\pi_i \circ b(x))_i \times (\tau'_{Sf,Tf}(f)_x)_f \\ &= (\pi_i \circ b(x))_i \times ((\tau_{Sf,Tf}(f) * i_b)_x)_f \\ &= (b(x)_i(*))_i \times (\tau_{Sf,Tf}(f)_{b(x)})_f \\ &= (b(x)_i(*))_i \times (\tau^{b(x)}_{Sf,Tf}(f)_*)_f \text{ by definition} \\ &= b(x) \text{ by the identification.}\end{aligned}$$

For $g : x \to y$ in $V$ we have

$$\begin{aligned}\psi \circ \phi(b)(g) &= \psi(\pi \circ b)(g) \\ &= (\pi_i \circ b(g))_i \\ &= (b(g)_i(*))_i \\ &= b(g) \text{ by the identification.}\end{aligned}$$

Thus $\psi \circ \phi(b)$ and $b$ agree as functors.

Next we verify the identity on morphisms. Let $\gamma : b \Rightarrow b'$ be a natural transformation. Then for $x \in V$ we have

$$\begin{aligned}\psi \circ \phi(\gamma)_x &= \psi(i_\pi * \gamma)_x \\ &= ((i_{\pi_j} * \gamma)_x)_j \\ &= (\pi_j(\gamma_x))_j \\ &= (\gamma_{xj}(*))_j \\ &= \gamma_x \text{ by the identification.}\end{aligned}$$

Thus $\psi \circ \phi(\gamma)$ and $\gamma$ agree as natural transformations and $\psi \circ \phi = 1_{Mor_{\mathcal{C}}(V,L)}$.

Another way to see this is to notice that $\pi_i$ is the projection onto the $i$-th coordinate. □

LEMMA 5.13. *The functor $\psi$ is a right inverse for $\phi$, i.e. $\phi \circ \psi = 1_{PseudoCone(V,F)}$.*

*Proof:* First we verify the identity on objects. Let $\pi' : \Delta_V \Rrightarrow F$ be a pseudo cone. For $j \in Obj\ \mathcal{J}$ and $x \in V$ we have

$$\begin{aligned}(\phi \circ \psi(\pi'))_j(x) &= (\pi \circ \psi(\pi'))_j(x)\\ &= \pi_j \circ \psi(\pi')(x)\\ &= \pi_j((\pi'_i(x))_i \times (\tau'_{Sf,Tf}(f)_x)_f)\\ &= \pi'_j(x).\end{aligned}$$

The last equality follows because $\pi_j$ is basically projection onto the $j$-th coordinate under the identification.

Next we verify the identity on morphisms. Let $\Xi : \alpha \rightsquigarrow \beta$ be a modification in $PseudoCone(V,F)$. For $j \in Obj\ \mathcal{J}$ and $x \in V$ we have

$$\begin{aligned}(\phi \circ \psi(\Xi))_j(x) &= (i_{\pi_j} * \psi(\Xi))_x\\ &= \pi_j(\psi(\Xi)_x)\\ &= \pi_j((\Xi_i(x))_i)\\ &= \Xi_j(x).\end{aligned}$$

Thus $\phi \circ \psi(\Xi) = \Xi$ and $\phi \circ \psi = 1_{PseudoCone(V,F)}$. □

LEMMA 5.14. *The small category $L$ with the pseudo cone $\pi : \Delta_L \Rrightarrow F$ is a pseudo limit of the pseudo functor $F : \mathcal{J} \to \mathcal{C}$.*

*Proof:* The functor $\phi : Mor_{\mathcal{C}}(V,L) \to PseudoCone(V,F)$ is an isomorphism of categories by the previous lemmas. Since $V$ was arbitrary we conclude that $L$ and $\pi$ are universal. □

Thus every pseudo functor $F : \mathcal{J} \to \mathcal{C}$ from a small 1-category $\mathcal{J}$ to the 2-category $\mathcal{C}$ of small categories admits a pseudo limit. In other words, the 2-category $\mathcal{C}$ of small categories admits pseudo limits. This completes the proof of Theorem 5.1. □

LEMMA 5.15. *The 2-category $\mathcal{C}$ of small categories admits cotensor products.*

*Proof:* Let $J$ and $F$ be small categories. Then $\{J,F\} := \mathcal{C}(J,F)$ is a cotensor product of $J$ and $F$ with unit $\pi : J \to \mathcal{C}(\mathcal{C}(J,F),F)$ defined by evaluation. □

THEOREM 5.16. *The 2-category $\mathcal{C}$ of small categories admits weighted pseudo limits.*

*Proof:* This 2-category admits 2-products. It also admits cotensor products and pseudo equalizers by Lemma 5.15 and Theorem 5.1. Theorem 3.22 then implies that it admits weighted pseudo limits. □

REMARK 5.17. The 2-category of small groupoids admits weighted pseudo limits.

*Proof:* The proof is exactly the same as the proof for small categories, since $L = PseudoCone(\mathbf{1}, F)$ is obviously a groupoid when the target of $F$ is the 2-category of small groupoids. □

THEOREM 5.18. *The 2-category of small categories and the 2-category of small groupoids admit weighted bilimits.*

*Proof:* They admit weighted pseudo limits, hence they also admit weighted bilimits. □

CHAPTER 6

# Theories and Algebras

The axioms for a group provide an example for the concept of a *theory* and an example of a group is an *algebra* over the theory of groups. In this chapter we describe what this means. Hu and Kriz point out in [**25**] that Lawvere's notion of a theory [**34**] is equivalent to another notion of theory. We prove this equivalence. It is well known that the category of algebras over a theory $T$ is equivalent to the category of algebras for some monad $C$. We present a version of this. Next we generalize theories in two ways: theories on a set of objects and theories enriched in groupoids. Theories on a set of objects allow us to describe algebraic structures on more than one set, such as modules or theories themselves. They also allow us to describe the free theory on a sequence of sets. Theories enriched in groupoids will be used in Chapter 7 to describe pseudo algebras over a theory $T$ as strict algebras over a theory $\mathcal{T}$ enriched in groupoids.

A theory can also be described as a finitary monad on the category $Sets$ of small sets as put forth in [**7**]. Theories on more than one object are called *many-sorted* in the monad description. Free finitary monads in the enriched and many-sorted contexts can be found in [**30**] and [**32**]. See [**47**] for monads in a general 2-category.

Definition 6.1. A *theory* is a category $T$ with objects $0, 1, 2, \ldots$ such that $n$ is the product of 1 with itself $n$ times in the category $T$ and each $n$ is equipped with a limiting cone.

This definition means for each $n \in Obj\ T$ we have chosen morphisms $pr_i : n \to 1$ for $i = 1, \ldots, n$ with the universal property: for any object $m \in Obj\ T$ and morphisms $w_i : m \to 1$ for $i = 1, \ldots, n$ there exists a unique morphism $\prod_{j=1}^n w_j : m \to n$ such that the diagram

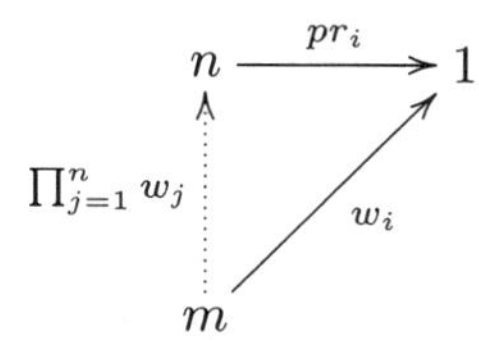


commutes for all $i = 1, \ldots, n$. In particular 0 is the terminal object of the category $T$. Note that we do not require the projection $pr_1 : 1 \to 1$ to be the identity, although it will automatically be an isomorphism. A useful notation is $T(n) := Mor_T(n, 1)$ for $n \in Obj\ T$. Elements of $T(n)$ are called *words of arity* $n$.

Another relevant morphism is the following. Let $\iota_i : \{1, \ldots, n_i\} \to \{1, \ldots, n_1 + n_2 + \cdots + n_k\}$ be the injective map which takes the domain to the $i$-th block and suppose that $w_i : n_i \to 1$ is a morphism for all $i = 1, \ldots k$. Then there exists a unique map denoted $(w_1, \dots, w_k)$ such that

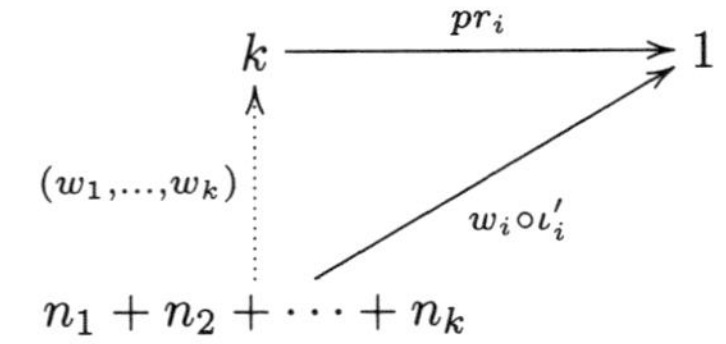


commutes for all $i = 1, \dots, k$ where $\iota'_i : n_1 + n_2 + \cdots + n_k \to n_i$ is the unique morphism such that

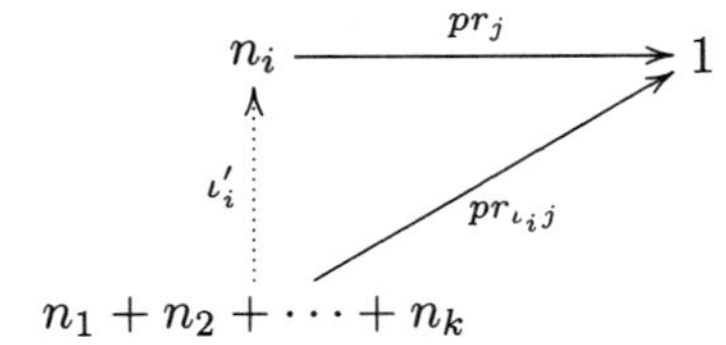


commutes. One should keep in mind that $n_1 + n_2 + \cdots + n_k$ is the product of $n_1, \dots, n_k$. Note that $(w_1, \dots, w_k)$ is not the same thing as the tuple $w_1, \dots, w_k$. The arrow $(w_1, \dots, w_k)$ is not the product of $w_1, \dots, w_k$.

LEMMA 6.2. *Let $T$ be a theory. Then $Mor_T(m, n)$ can be identified with the set-theoretic product $\prod_{j=1}^{n} Mor_T(m, 1)$ via the map which takes $w$ to the tuple with entries $pr_1 \circ w, \dots, pr_n \circ w$. We identify $w$ with that tuple. In particular a theory is determined up to isomorphism by the sets $T(0), T(1), T(2), \dots$.*

*Proof:* This follows directly from the definition of product in a category. □

EXAMPLE 6.3. Let $X$ be a set. Then the *endomorphism theory End(X)* has objects $0, 1, 2, \dots$ and hom sets $Mor_{End(X)}(m, n) = Map(X^m, X^n)$. Composition is the usual function composition. Here we readily see that $\{*\}$ is the terminal object and that $End(X)(0) = Mor_{End(X)}(0, 1)$ can be identified with $X$.

Let $w \in End(X)(k)$ and $w_i \in End(X)(n_i)$ for $i = 1, \dots, k$. Then the composite function $\gamma(w, w_1, \dots, w_k) := w \circ (w_1, \dots, w_k)$ is an element of $End(X)(n_1 + \cdots + n_k)$. This *composition* is associative. Let $1 := 1_X \in End(X)(1)$. Then apparently $w \circ (1, \dots, 1) = w$ and $1 \circ w = w$, *i.e.* the composition is also unital.

Let $\{1, \dots, k\} \xrightarrow{f} \{1, \dots, \ell\} \xrightarrow{g} \{1, \dots, m\}$ be maps of sets. For a word $w \in End(X)(k)$ we define a new word $w_f \in End(X)(\ell)$ by $w_f(x_1, \dots, x_\ell) := w(x_{f1}, \dots, x_{fk})$ called the *substituted word*. Thus we have maps

$$End(X)(k) \xrightarrow{()_f} End(X)(\ell) \xrightarrow{()_g} End(X)(m).$$

If $e : \emptyset \to \{1, \dots, k\}$ is the empty function and $x \in X = End(X)(0)$, then the substituted word $x_e : X^k \to X$ is the constant function $(x_1, \dots, x_k) \mapsto x$. There are no other functions $\emptyset \to \{1, \dots, k\}$. We easily see that $(w_f)_g = w_{g \circ f}$ and $w_{id_k} = w$ for the identity map $id_k : \{1, \dots, k\} \to \{1, \dots, k\}$, *i.e.* these *substitution maps* are functorial.

These substitution maps relate to the composition in two ways, which we now describe. Let $f : \{1, \dots, k\} \to \{1, \dots, \ell\}$, $w \in End(X)(k)$, and $w_i \in End(X)(n_i)$ for $i = 1, \dots, \ell$. Then $w_f \circ (w_1, \dots, w_\ell) = (w \circ (w_{f1}, \dots, w_{fk}))_{\bar{f}}$ where

$$\bar{f} : \{1, 2, \dots, n_{f1} + n_{f2} + \cdots + n_{fk}\} \longrightarrow \{1, 2, \dots, n_1 + n_2 + \cdots + n_\ell\}$$

is the function obtained by parsing the sequence $1, 2, \ldots, n_1 + n_2 + \cdots + n_\ell$ into consecutive blocks $B_1, \ldots, B_\ell$ of lengths $n_1, \ldots, n_\ell$ respectively and then writing them in the order $B_{f1}, \ldots, B_{fk}$. For example, let $n_1 = 1, n_2 = 2, n_3 = 3, n_4 = 1, w \in T(3)$, and $w_i \in T(n_i)$ for $i = 1, \ldots, 4$ and let $f : \{1,2,3\} \to \{1,2,3,4\}$ be given by

$$\begin{pmatrix} 1 & 2 & 3 \\ 3 & 2 & 4 \end{pmatrix}.$$

Then $\bar{f} : \{1, 2, \ldots, 6\} \to \{1, 2, \ldots, 7\}$ is given by

$$\begin{pmatrix} 1 & 2 & 3 & 4 & 5 & 6 \\ B_{f1} & & & B_{f2} & & B_{f3} \end{pmatrix} = \begin{pmatrix} 1 & 2 & 3 & 4 & 5 & 6 \\ 4 & 5 & 6 & 2 & 3 & 7 \end{pmatrix}.$$

We see that

$$\begin{aligned} w_f \circ (w_1, w_2, w_3, w_4)(x_1, \ldots, x_7) &= w_f(w_1(x_1), w_2(x_2, x_3), w_3(x_4, x_5, x_6), w_4(x_7)) \\ &= w(w_3(x_4, x_5, x_6), w_2(x_2, x_3), w_4(x_7)) \\ &= w \circ (w_{f1}, w_{f2}, w_{f3})(x_4, x_5, x_6, x_2, x_3, x_7) \\ &= w \circ (w_{f1}, w_{f2}, w_{f3})(x_{\bar{f}1}, x_{\bar{f}2}, \ldots, x_{\bar{f}6}). \end{aligned}$$

In other words we have $w_f \circ (w_1, w_2, w_3) = (w \circ (w_{f1}, w_{f2}, w_{f3}))_{\bar{f}}$. Note that $\bar{f}$ depends not only on $f$, but also on the arity of the words we are composing. The equality $w_f \circ (w_1, \ldots, w_\ell) = (w \circ (w_{f1}, \ldots, w_{fk}))_{\bar{f}}$ is the first relationship between composition and the substitution maps $()_f$.

The second way the composition and the substitution maps relate occurs in the following situation. If $w \in End(X)(k)$, $w_i \in End(X)(n_i)$, and $g_i : \{1, \ldots, n_i\} \to \{1, \ldots, n_i'\}$ are functions for $i = 1, \ldots, k$, then $w \circ ((w_1)_{g_1}, \ldots, (w_k)_{g_k}) = (w \circ (w_1, \ldots, w_k))_{g_1 + \cdots + g_k}$ where $g_1 + \cdots + g_k : \{1, 2, \ldots, n_1 + \cdots + n_k\} \to \{1, 2, \ldots, n_1' + \cdots + n_k'\}$ is the function obtained by placing $g_1, \ldots, g_k$ next to each other from left to right.

EXAMPLE 6.4. Let $X$ be a category. Then the *endomorphism theory End(X)* has objects $0, 1, 2, \ldots$ and it has hom sets $Mor_{End(X)}(m, n) = Functors(X^m, X^n)$. We can proceed as in the previous example and define substituted functors (substituted words). Note that $End(X)$ can be made into a 2-category by taking the 2-cells to be natural transformations, although we leave out the 2-cells for now. In most applications we will only be concerned with the 1-category $End(X)$.

EXAMPLE 6.5. Let $X$ be an object of a category with finite products. Then we obtain a theory $End(X)$ with hom sets $Mor_{End(X)}(m, n) := Mor(X^m, X^n)$.

We can abstract the essential properties of $End(X)$ in the previous examples to get the following lemma for arbitrary theories.

LEMMA 6.6. *Let $T$ be a theory. Then for all $k, n_1, \ldots, n_k \in \{0, 1, \ldots\}$ there is a map $\gamma : T(k) \times T(n_1) \times \cdots \times T(n_k) \to T(n_1 + \cdots + n_k)$ called composition and for every function $f : \{1, \ldots, k\} \to \{1, \ldots, \ell\}$ there is a map $T(k) \xrightarrow{()_f} T(\ell)$ called substitution. These maps have the following properties.*

(1) *The $\gamma$'s are associative, i.e.*

$$\gamma(w, \gamma(w^1, w_1^1, \ldots, w_{n_1}^1), \gamma(w^2, w_1^2, \ldots, w_{n_2}^2), \ldots, \gamma(w^k, w_1^k, \ldots, w_{n_k}^k)) = \\ \gamma(\gamma(w, w^1, \ldots, w^k), w_1^1, \ldots, w_{n_1}^1, w_1^2, \ldots, w_{n_2}^2, \ldots, w_1^k, \ldots, w_{n_k}^k).$$

(2) *The $\gamma$'s are unital, i.e. there exists an element $1 \in T(1)$ called the unit such that*

$$\gamma(w, 1, \ldots, 1) = w = \gamma(1, w)$$

*for all $w \in T(k)$. Moreover, such an element is unique.*

(3) *The $\gamma$'s are equivariant in the sense that*

$$\gamma(w_f, w_1, \ldots, w_\ell) = \gamma(w, w_{f1}, \ldots, w_{fk})_{\bar{f}}$$

*for all $f : \{1, \ldots, k\} \to \{1, \ldots, \ell\}$ where $\bar{f} : \{1, 2, \ldots, n_{f1} + n_{f2} + \cdots + n_{fk}\} \to \{1, 2, \ldots, n_1 + n_2 + \cdots + n_\ell\}$ is the function that moves entire blocks according to $f$ as mentioned in the example above. Here $\bar{f}$ depends also on the particular $\gamma$.*

(4) *The $\gamma$'s are equivariant in the sense that*

$$\gamma(w, (w_1)_{g_1}, \ldots, (w_k)_{g_k}) = \gamma(w, w_1, \ldots, w_k)_{g_1 + \cdots + g_k}$$

*for all functions $g_i : \{1, \ldots, n_i\} \to \{1, \ldots, n_i'\}$ where $g_1 + \cdots + g_k : \{1, 2, \ldots, n_1 + \cdots + n_k\} \to \{1, 2, \ldots, n_1' + \cdots + n_k'\}$ is the function obtained by placing $g_1, \ldots, g_k$ next to each other from left to right.*

(5) *The substitution is functorial, i.e. for functions $\{1, \ldots, k\} \xrightarrow{f} \{1, \ldots, \ell\} \xrightarrow{g} \{1, \ldots, m\}$ the composition*

$$T(k) \xrightarrow{()_f} T(\ell) \xrightarrow{()_g} T(m)$$

*is the same as*

$$T(k) \xrightarrow{()_{g \circ f}} T(m)$$

*and for the identity function $id_k : \{1, \ldots, k\} \to \{1, \ldots, k\}$ the map*

$$T(k) \xrightarrow{()_{id_k}} T(k)$$

*is equal to the identity for all $k \geq 0$.*

*Proof:* First we define the substitution. Let $f : \{1, \ldots, k\} \to \{1, \ldots, \ell\}$ be a function. Then there exists a unique morphism $f'$ such that the diagram

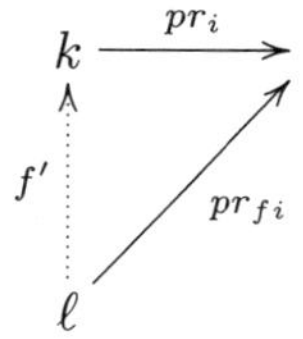


commutes for all $i = 1, \ldots, k$. For $w \in T(k)$ define $w_f := w \circ f'$. Thus the map $T(k) \xrightarrow{()_f} T(\ell)$ is defined by precomposition with $f'$.

Next we define the composition $\gamma : T(k) \times T(n_1) \times \cdots \times T(n_k) \to T(n_1 + n_2 + \cdots + n_k)$. Let $w \in T(k), w_i \in T(n_i)$ for $i = 1, \ldots, k$. Define $\gamma(w, w_1, \ldots, w_k) := w \circ (w_1, \ldots, w_k)$ where the composition $\circ$ is the composition of the category $T$ and

$(w_1, \dots, w_k)$ is the unique morphism such that

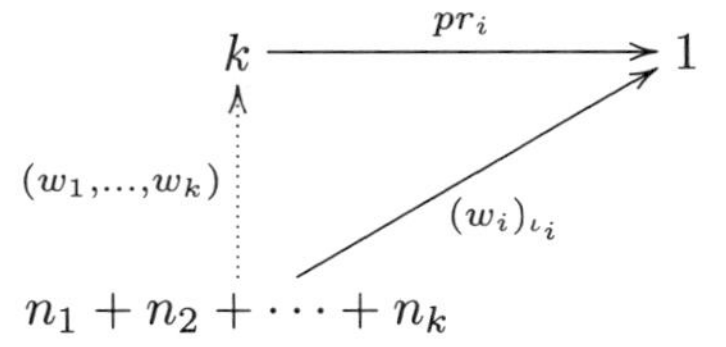


commutes as defined above.

(1) We claim that $\gamma$ is associative.
$\gamma(w, \gamma(w^1, w^1_1, \dots, w^1_{n_1}), \gamma(w^2, w^2_1, \dots, w^2_{n_2}), \dots, \gamma(w^k, w^k_1, \dots, w^k_{n_k})) =$
$= w \circ (w^1 \circ (w^1_1, \dots, w^1_{n_1}), \dots, w^k \circ (w^k_1, \dots, w^k_{n_k}))$
$= w \circ ((w^1, \dots, w^k) \circ ((w^1_1, \dots, w^1_{n_1}), \dots, (w^k_1, \dots, w^k_{n_k})))$
$= (w \circ (w^1, \dots, w^k)) \circ (w^1_1, \dots, w^1_{n_1}, \dots, w^k_1, \dots, w^k_{n_k})$
$= \gamma(\gamma(w, w^1, \dots, w^k), w^1_1, \dots, w^1_{n_1}, w^2_1, \dots, w^2_{n_2}, \dots, w^k_1, \dots, w^k_{n_k})$
The second to last equality follows by associativity of composition in the category $T$ and by properties of products.

(2) We claim that $\gamma$ is unital. Let $1 : 1 \to 1$ be the projection morphism of the object 1 in the category $T$, which is not necessarily the identity morphism of the object 1. Then $(1, \dots, 1) : k \to k$ is the identity morphism of the object $k$ because $1_{\iota_i} = 1 \circ \iota'_i = 1 \circ (pr_1^{-1} \circ pr_i) = pr_1 \circ (pr_1^{-1} \circ pr_i) = pr_i$ in the diagram

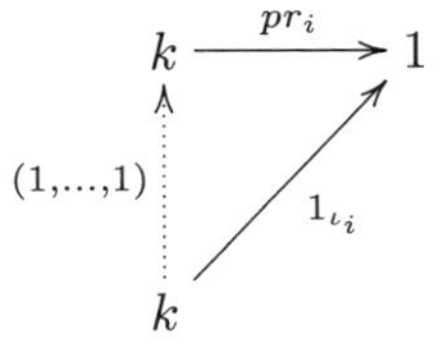


for all $i = 1, \dots, k$. Here $\iota_i : \{1\} \to \{1, \dots, k\}$ is defined by $\iota_i(1) = i$. Thus $\gamma(w, 1, \dots, 1) = w \circ (1, \dots, 1) = w \circ 1_k = w$.

To show $\gamma(1, w) = w$ we consider the diagram

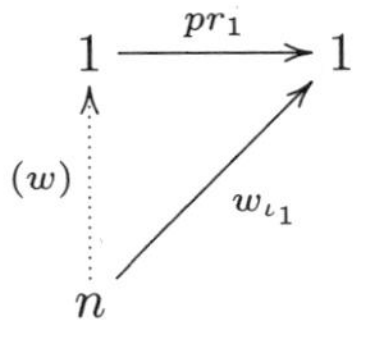


where $\iota_1 : \{1, \dots, n\} \to \{1, \dots, n\}$ is the identity. Then $w_{\iota_1} = w$ and $(w) = pr_1^{-1} \circ w$. Thus $\gamma(1, w) = 1 \circ (w) = pr_1 \circ (pr_1^{-1} \circ w) = w$.

The uniqueness follows from $1 = \gamma(1, 1') = 1'$.

(3) Let $f : \{1, \dots, k\} \to \{1, \dots, \ell\}$ be a function and $w_i \in T(i)$ for $i = 1, \dots, \ell$. Using the definitions of $\bar{f} : \{1, \dots, n_{f1} + \cdots + n_{fk}\} \to \{1, \dots, n_1 + \cdots + n_\ell\}$

and $\iota_i$ from above we see that the following two diagrams

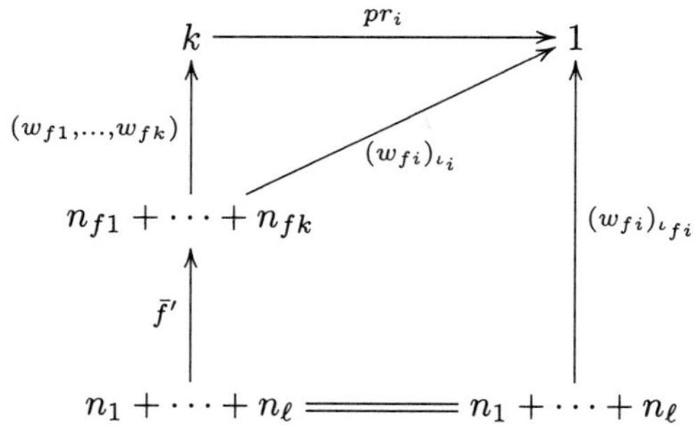


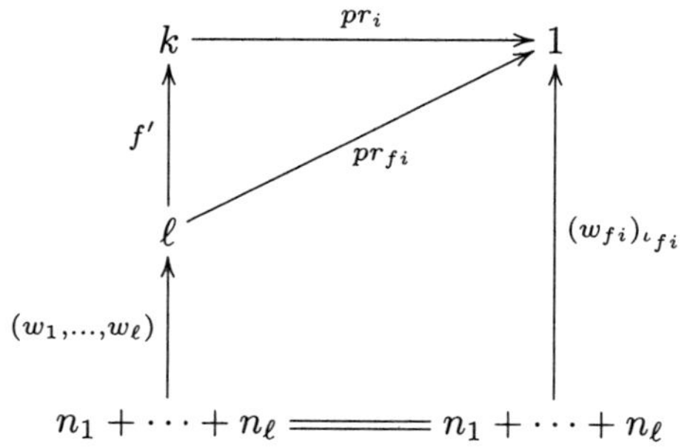


commute for all $i = 1, \ldots, k$. Hence by the universal property of the product $k$ we have $f' \circ (w_1, \ldots, w_\ell) = (w_{f1}, \ldots, w_{fk}) \circ \bar{f}'$. Using this we see that

$$\begin{aligned} \gamma(w_f, w_1, \ldots, w_\ell) &= w \circ f' \circ (w_1, \ldots, w_\ell) \\ &= w \circ (w_{f1}, \ldots, w_{fk}) \circ \bar{f}' \\ &= \gamma(w, w_{f1}, \ldots, w_{fk})_{\bar{f}}. \end{aligned}$$

(4) Let $g_i : \{1, \ldots, n_i\} \to \{1, \ldots, n_i'\}$ be functions for $i = 1, \ldots, k$. Then

$$\begin{aligned} \gamma(w, (w_1)_{g_1}, \ldots, (w_k)_{g_k}) &= w \circ (w_1 \circ g_1', \ldots, w_k \circ g_k') \\ &= w \circ (w_1, \ldots, w_k) \circ (g_1', \ldots, g_k') \\ &= w \circ (w_1, \ldots, w_k) \circ (g_1 + \cdots + g_k)' \\ &= \gamma(w, w_1, \ldots, w_k)_{g_1 + \cdots + g_k}. \end{aligned}$$

(5) Let $\{1, \ldots, k\} \xrightarrow{f} \{1, \ldots, \ell\} \xrightarrow{g} \{1, \ldots, m\}$ be functions. Then $f'$ and $g'$ make the two small subdiagrams in

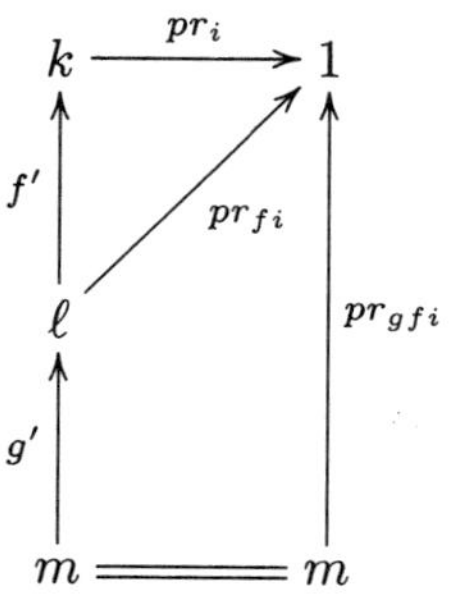

commute for all $i = 1, \ldots, k$. Thus the outer diagram commutes and $(g \circ f)' = f' \circ g'$ by the universal property of the product. We conclude $(w_f)_g = w \circ f' \circ g' = w \circ (g \circ f)' = w_{g \circ f}$. The identity $1_k : k \to k$ makes

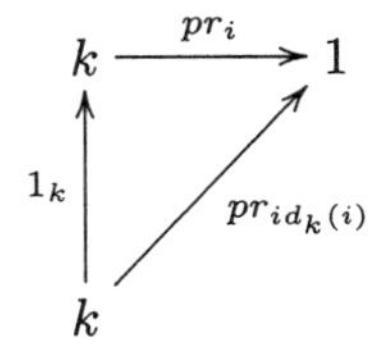


commute for all $i = 1, \ldots, k$ where $id_k : \{1, \ldots, k\} \to \{1, \ldots, k\}$ is the identity function. Hence $(id_k)' = 1_k$ and $w_{id_k} = w \circ (id_k)' = w \circ 1_k = w$ for all $w \in T(k)$.

We have verified all of the axioms. □

There is another description of a theory which can be formulated by using the category $\Gamma$.

DEFINITION 6.7. Let $\Gamma$ be the category with objects $\emptyset = 0, 1, 2, \ldots$ where $k = \{1, \ldots, k\}$. The morphisms $k \to \ell$ are just maps of sets. In particular 0 is the initial object since the only map $\emptyset \to k$ is the empty function. There are no maps $k \to \emptyset$ for $k \geq 1$. The object 1 is the terminal object. Let $+ : \Gamma \times \Gamma \to \Gamma$ denote the usual functor obtained by adding the sets and placing maps side by side.

REMARK 6.8. Let $T$ be a theory. Then by the previous lemma $T$ defines a functor from $\Gamma$ to *Sets* by $k \mapsto T(k)$ and $f \mapsto ()_f$. Moreover, this functor comes with maps $\gamma : T(k) \times T(n_1) \times \cdots \times T(n_k) \to T(n_1 + \cdots + n_k)$ which satisfy 1. through 5. The compositions $\gamma$, unit 1, and substitution are sometimes called the *operations of theories*. The relations in 1. through 5. are sometimes called the *relations of theories*.

LEMMA 6.9. *Let $T$ be a functor from $\Gamma$ to Sets equipped with maps $\gamma : T(k) \times T(n_1) \times \cdots \times T(n_k) \to T(n_1 + \cdots + n_k)$ and an element $1 \in T(1)$ which satisfy (1) through (5) where $T(f) =: ()_f$ for functions $f : k \to \ell$. Then $T$ determines a theory with $Mor(n, 1) = T(n)$ for all $n \geq 0$.*

*Proof:* Define the underlying category of the theory to formally have objects $0, 1, 2 \ldots$ and morphisms $Mor(m, n) := \prod_{i=1}^{n} Mor(m, 1)$. In particular $Mor(m, 0)$ only has one element. We denote a tuple of words $w_1, \ldots, w_n \in Mor(m, 1)$ by $\prod_{i=1}^{n} w_i$. For $k, \ell \geq 0$ let $\iota_{\ell,k} : \{1, \ldots, \ell k\} \to \{1, \ldots, k\}$ be the function such that $\iota_{\ell,k}(i + jk) = i$ for $i = 1, \ldots, k$, in other words $\iota_{\ell,k}$ wraps the domain around the codomain $\ell$ times. Now define the composition of $\prod_{i=1}^{\ell} w_i \in Mor(k, \ell)$ with $\prod_{i=1}^{m} v_i \in Mor(\ell, m)$ to be $\prod_{i=1}^{m} v_i \circ \prod_{i=1}^{\ell} w_i := \prod_{i=1}^{m} \gamma(v_i, w_1, \ldots, w_\ell)_{\iota_{\ell,k}}$. This composition is associative because $\gamma$ is associative and equivariant.

Let $f_i : \{1\} \to \{1, \ldots, n\}$ be the map $f_i(1) = i$. Define $pr_i := 1_{f_i} \in T(n)$ where $1 \in T(1)$ is the distinguished element whose existence we assumed. This notation is slightly imprecise because we have different sequences $pr_1, \ldots, pr_n$ for different $n \geq 0$. From the context it will always be clear which sequence of morphisms is meant. We claim that $\prod_{i=1}^{n} pr_i \in Mor(n, n)$ is the identity on the object $n$. Let

$\prod_{i=1}^m w_i \in Mor(n, m)$. Then

$$\begin{aligned}
\prod_{i=1}^m w_i \circ \prod_{i=1}^n pr_i &= \prod_{i=1}^m \gamma(w_i, pr_1, \ldots, pr_n)_{\iota_{n,n}} \\
&= \prod_{i=1}^m \gamma(w_i, 1_{f_1}, \ldots, 1_{f_n})_{\iota_{n,n}} \\
&= \prod_{i=1}^m (\gamma(w_i, 1, \ldots, 1)_{f_1+\cdots+f_n})_{\iota_{n,n}} \text{ by equivariance} \\
&= \prod_{i=1}^m \gamma(w_i, 1, \ldots, 1)_{\iota_{n,n}\circ(f_1+\cdots+f_n)} \text{ by functoriality of } T \\
&= \prod_{i=1}^m w_i \text{ since } \gamma \text{ is unital, } \iota_{n,n} \circ (f_1 + \cdots + f_n) = id_n, \\
&\qquad \text{and functoriality of } T.
\end{aligned}$$

Now for the other side let $\prod_{i=1}^n w_i \in Mor(m, n)$. Then

$$\begin{aligned}
\prod_{i=1}^n pr_i \circ \prod_{i=1}^n w_i &= \prod_{i=1}^n \gamma(pr_i, w_1, \ldots, w_n)_{\iota_{n,m}} \text{ by definition} \\
&= \prod_{i=1}^n \gamma(1_{f_i}, w_1, \ldots, w_n)_{\iota_{n,m}} \text{ by definition} \\
&= \prod_{i=1}^n (\gamma(1, w_i)_{\bar{f}_i})_{\iota_{n,m}} \text{ by equivariance} \\
&= \prod_{i=1}^n (w_i)_{\iota_{n,m}\circ\bar{f}_i} \text{ by unitality of } \gamma \text{ and functoriality of } T \\
&= \prod_{i=1}^n w_i \text{ since } \iota_{n,m} \circ \bar{f}_i = id_m.
\end{aligned}$$

This can be seen by observing that $\bar{f}_i : \{1, \ldots, m\} \to \{1, \ldots, nm\}$ has the form

$$\begin{pmatrix} 1 & 2 & \ldots & m \\ (i-1)m+1 & (i-1)m+2 & \ldots & (i-1)m+m \end{pmatrix}$$

and by using the definition of $\iota_{n,m}$. Thus $\prod_{i=1}^n pr_i \in Mor(n, n)$ is the identity on the object $n$.

Thus far we have shown that we have a category with objects $0, 1, 2, \ldots$ and morphisms $Mor(m, n)$. We claim that $n$ is the product of $n$ copies of 1 in this category with projections $pr_1, \ldots, pr_n : n \to 1$ introduced above. First note for $\prod_{i=1}^n w_i \in Mor(m, n)$ we have

$$\begin{aligned}
pr_i \circ \prod_{i=1}^n w_i &= \gamma(pr_i, w_1, \ldots, w_n)_{\iota_{n,m}} \\
&= \gamma(1_{f_i}, w_1, \ldots, w_n)_{\iota_{n,m}} \text{ by definition} \\
&= \gamma(1, w_i)_{\iota_{n,m}\circ\bar{f}_i} \text{ by equivariance and functoriality} \\
&= w_i \text{ since } \iota_{n,m} \circ \bar{f}_i = id_m \text{ and by functoriality.}
\end{aligned}$$

Now suppose we are given morphisms $w_1, \dots, w_n \in Mor(m, 1)$. Then

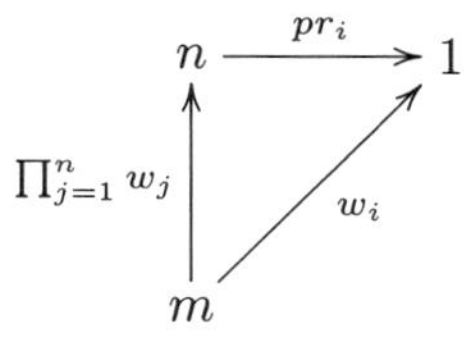


commutes for all $i = 1, \dots, n$ by the remark just made. If $\prod_{i=1}^n v_i \in Mor(m, n)$ is another morphism such that

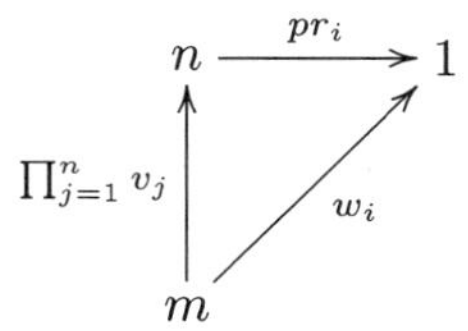


commutes for all $i = 1, \dots, n$, then by the remark $v_i = pr_i \circ \prod_{j=1}^n v_j = w_i$ and hence $\prod_{j=1}^n v_j = \prod_{j=1}^n w_j$ and the factorizing map is unique. Hence $n$ is the product of $n$ copies of 1.

We conclude that the functor $T$ with the maps $\gamma$ satisfying the axioms (1) through (5) determines a theory with the indicated hom sets. □

THEOREM 6.10. *A theory $T$ is determined by either of the following equivalent collections of data:*

1. *A category $T$ with objects $0, 1, 2, \dots$ such that $n$ is the categorical product of 1 with itself $n$ times and each $n$ is equipped with a choice of projections.*
2. *A functor $T : \Gamma \to Sets$ equipped with maps $\gamma : T(k) \times T(n_1) \times \cdots \times T(n_k) \to T(n_1 + \cdots + n_k)$ and a unit $1 \in T(1)$ which satisfy (1) through (5) of Lemma 6.6.*

*Proof:* In each description $Mor_T(n, 1)$ is the same. By the universality of products this determines the rest of the theory. The two processes of Lemmas 6.6 and 6.9 are "inverse" to one another by further inspection, provided we identify $Mor_T(m, n)$ with $\prod_{i=1}^n T(m)$. □

DEFINITION 6.11. Let $S$ and $T$ be theories. In the categorical description of $S$ and $T$ a *morphism of theories* $\Phi : S \to T$ is a functor from the category $S$ to the category $T$ such that $\Phi(n_S) = n_T$ and $\Phi(pr_i) = pr_i$ for all projections.

One easily sees that the theories form a category and we have a suitable forgetful functor.

THEOREM 6.12. *The forgetful functor from the category of theories to $\prod_{n \geq 0} Sets$ given by $T \mapsto (T(0), T(1), \dots)$ admits a left adjoint called the free theory functor.*

*Proof:* On page 56 we will construct the free theory on the sequence of sets $(T(0), T(1), \dots)$. □

To make later proofs easier, we need the following lemma.

LEMMA 6.13. *Let $\Phi : S \to T$ be a morphism of theories.*

(1) *Let $f : \{1, \ldots, k\} \to \{1, \ldots, \ell\}$ be a function. As usual, $f' : \ell \to k$ denotes the unique morphism in any theory such that*

$$\begin{array}{ccc} k & \xrightarrow{pr_i} & 1 \\ {\scriptstyle f'}\uparrow & \nearrow{\scriptstyle pr_{fi}} & \\ \ell & & \end{array}$$

*commutes. Then $\Phi(f') = f'$.*

(2) *Let $f : \{1, \ldots, k\} \to \{1, \ldots, \ell\}$ be a function and $w \in Mor_S(k, 1)$. Then $\Phi(w_f) = \Phi(w)_f$.*

(3) *Let $w_1, \ldots, w_n \in Mor_S(m, 1)$. Then $\Phi(\prod_{j=1}^n w_j) = \prod_{j=1}^n \Phi(w_j)$.*

(4) *Let $w_i \in Mor_S(n_i, 1)$ for $i = 1, \ldots, k$. Then we also have $\Phi(w_1, \ldots, w_k) = (\Phi(w_1), \ldots, \Phi(w_k))$.*

*Proof:*

(1) The diagram

$$\begin{array}{ccc} k & \xrightarrow{pr_i} & 1 \\ {\scriptstyle \Phi(f')}\uparrow & \nearrow{\scriptstyle pr_{fi}} & \\ \ell & & \end{array}$$

commutes for all $i = 1, \ldots, k$ by the properties of $\Phi$. Then $\Phi(f') = f'$ by the universal property of the product.

(2) This follows from (1) and the definition $w_f = w \circ f'$.

(3) The properties of $\Phi$ imply that the diagram

$$\begin{array}{ccc} n & \xrightarrow{pr_i} & 1 \\ {\scriptstyle \Phi(\prod_{j=1}^n w_j)}\uparrow & \nearrow{\scriptstyle \Phi(w_i)} & \\ m & & \end{array}$$

commutes for all $i = 1, \ldots, n$. Then $\Phi(\prod_{j=1}^n w_j) = \prod_{j=1}^n \Phi(w_j)$ by the universal property of the product.

(4) By (2) we have $\Phi((w_i)_{\iota_i}) = \Phi(w_i)_{\iota_i}$. Hence, the properties of $\Phi$ imply that the diagram

$$\begin{array}{ccc} k & \xrightarrow{pr_i} & 1 \\ {\scriptstyle \Phi(w_1, \ldots, w_k)}\uparrow & \nearrow{\scriptstyle \Phi(w_i)_{\iota_i}} & \\ n_1 + n_2 + \cdots + n_k & & \end{array}$$

commutes for all $i = 1, \ldots, k$. Then $\Phi(w_1, \ldots, w_k) = (\Phi(w_1), \ldots, \Phi(w_k))$ by the universal property of the product.

□

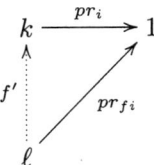

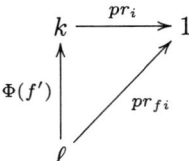

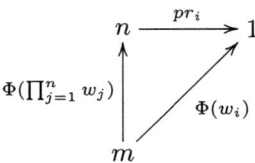

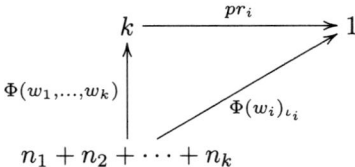

Just as a theory has a categorical description and a functorial description, a morphism of theories also has a second description. We work towards the second description in the following two lemmas.

LEMMA 6.14. *Let $\Phi : S \to T$ be a morphism of theories, i.e. a functor such that $\Phi(n_S) = n_T$ and $\Phi(pr_i) = pr_i$ for all projections. Then $\Phi$ determines a natural transformation $S \Rightarrow T$ also denoted by $\Phi$ such that*

$$\begin{CD}
S(k) \times S(n_1) \times \cdots \times S(n_k) @>{\Phi_k \times \Phi_{n_1} \times \cdots \times \Phi_{n_k}}>> T(k) \times T(n_1) \times \cdots \times T(n_k) \\
@V{\gamma^S}VV @VV{\gamma^T}V \\
S(n_1 + \cdots + n_k) @>>{\Phi_{n_1+\cdots+n_k}}> T(n_1 + \cdots + n_k)
\end{CD}$$

*commutes and $\Phi_1(1_S) = 1_T$, where $S, T : \Gamma \to Sets$ are the functors in the functorial description of the theories $S$ and $T$.*

*Proof:* Let $\Phi_m : Mor_S(m, 1) \to Mor_T(m, 1)$ denote the map obtained from the functor $\Phi$, *i.e.* $\Phi_m(w) := \Phi(w)$ for $w \in S(m)$. Then for $f : m \to n$ in $\Gamma$ and $w \in S(m)$, we have $\Phi(w_f) = \Phi(w)_f$ by Lemma 6.13. Hence

$$\begin{CD}
S(m) @>{\Phi_m}>> T(m) \\
@V{S(f)}VV @VV{T(f)}V \\
S(n) @>>{\Phi_n}> T(n)
\end{CD}$$

commutes and $m \mapsto \Phi_m$ is natural.

Let $w \in S(k)$ and $w_i \in S(n_i)$ for $i = 1, \ldots, k$. Then

$$\begin{aligned}
\Phi_{n_1+\cdots+n_k}(\gamma^S(w, w_1, \ldots, w_k)) &= \Phi(w \circ (w_1, \ldots, w_k)) \\
&= \Phi(w) \circ (\Phi(w_1), \ldots, \Phi(w_k)) \\
&= \gamma^T(\Phi(w), \Phi(w_1), \ldots, \Phi(w_k)) \\
&= \gamma^T(\Phi_k(w), \Phi_{n_1}(w_1), \ldots, \Phi_{n_k}(w_k)).
\end{aligned}$$

Hence the natural transformation $m \mapsto \Phi_m$ preserves the $\gamma$'s.

Let $1_S \in S(1)$ and $1_T \in T(1)$ be the units in the respective theories. Then $\Phi_1(1_S) = 1_T$ because the functor $\Phi$ preserves projections.

Thus $\Phi : S \Rightarrow T$ is a natural transformation which preserves the compositions and the units. □

LEMMA 6.15. *Let $S, T : \Gamma \to Sets$ be theories. Let $\Phi : S \Rightarrow T$ be a natural transformation preserving the $\gamma$'s and their units as in Lemma 6.14. Then $\Phi$ determines a functor $S \to T$ also denoted $\Phi$, where $S$ and $T$ are the categories in the categorical description of the theories $S, T : \Gamma \to Sets$. Moreover, the functor $\Phi : S \to T$ satisfies $\Phi(n_S) = n_T$ and $\Phi(pr_i) = pr_i$ for all projections.*

*Proof:* We define $\Phi(n_S) = n_T$ for all $n_S \in Obj\ S$ and $\Phi(\prod_{j=1}^{\ell} w_j) := \prod_{j=1}^{\ell} \Phi_k(w_j)$ for all $\prod_{j=1}^{\ell} w_j \in Mor_S(k,\ell)$. Then for $\prod_{i=1}^{m} v_i \in Mor_S(\ell, m)$ we have

$$\begin{aligned}\Phi(\prod_{i=1}^{m} v_i \circ \prod_{j=1}^{\ell} w_j) &= \Phi(\prod_{i=1}^{m} \gamma(v_i, w_1, \ldots, w_\ell)_{\iota_{\ell,k}}) \text{ from Lemma 6.9}\\ &= \prod_{i=1}^{m} \gamma(\Phi_\ell(v_i), \Phi_k(w_1), \ldots, \Phi_k(w_\ell))_{\iota_{\ell,k}}\\ &= \prod_{i=1}^{m} \Phi_\ell(v_j) \circ \prod_{j=1}^{\ell} \Phi_k(w_j)\\ &= \Phi(\prod_{i=1}^{m} v_i) \circ \Phi(\prod_{j=1}^{\ell} w_j).\end{aligned}$$

Hence $\Phi$ preserves compositions.

We claim that $\Phi$ preserves projections. Let $f_i : \{1\} \to \{1, \ldots, n\}$ be the map $f_i(1) = i$. Then $(1_S)_{f_i} = pr_i$ and

$$\begin{aligned}\Phi(pr_i) &= \Phi_n((1_S)_{f_1})\\ &= \Phi_n(1_S)_{f_i} \text{ by naturality}\\ &= (1_T)_{f_i} \text{ since } \Phi \text{ preserves the unit}\\ &= pr_i.\end{aligned}$$

Hence $\Phi$ preserves projections.

We claim that $\Phi$ preserves identities. Recall that $\prod_{j=1}^{n} pr_j : n \to n$ is the identity on the object $n$ of the category $S$. Then

$$\begin{aligned}\Phi(\prod_{j=1}^{n} pr_j) &= \prod_{j=1}^{n} \Phi(pr_j) \text{ by definition}\\ &= \prod_{j=1}^{n} pr_j \text{ because } \Phi \text{ preserves projections.}\end{aligned}$$

Thus $\Phi$ preserves identities and is a functor $S \to T$. □

Combining these two lemmas gives us the two descriptions of a morphism of theories in the following theorem.

THEOREM 6.16. *Let $S$ and $T$ be theories. Then a morphism $S \to T$ of theories is given by either of the following equivalent collections of data:*

1. *A functor $\Phi : S \to T$ such that $\Phi(n_S) = n_T$ for all $n_S \in Obj\ S$ and $\Phi(pr_i) = pr_i$ for all projections.*
2. *A natural transformation $\Phi : S \Rightarrow T$ of the functors $S, T : \Gamma \to Sets$ which preserves the $\gamma$'s and the units.*

*Proof:* The processes of the previous two lemmas are "inverse" to each other by inspection. □

THEOREM 6.17. *The category of theories with objects and morphisms as in (1) of Theorems 6.10 and 6.16 is equivalent to the category with objects and morphisms as in (2) of Theorems 6.10 and 6.16.*

*Proof:* This relies on the bijection $Mor_T(m,n) \cong \prod_{j=1}^n Mor_T(m,1)$. $\square$

The concept of an *algebra* is closely related to the concept of theories. Roughly speaking, an algebra over a theory is a category together with a rule that assigns an $n$-ary operation on $X$ to every word of the theory of arity $n$ in such way that compositions, substitutions, and identity 1 are preserved.

DEFINITION 6.18. Let $X$ be a category and $T$ a theory. Then $X$ is a *$T$-algebra* if it is equipped with a morphism of theories $T \to End(X)$, where $End(X)$ is the theory in Example 6.4. We also say $X$ is an *algebra over the theory $T$*.

Notice that if $X$ is a set viewed as a discrete category, this is the usual definition of an algebra over a theory. Note also that we have two versions of $T$-algebra, one is given by the categorical description of theories and the other by the functorial description. A familiar example of an algebra is a group, since a group is an algebra over the theory of groups as follows.

EXAMPLE 6.19. Let $T$ be the theory of groups, *i.e.* there are morphisms $e \in T(0), \nu \in T(1)$, and $\mu \in T(2)$ which satisfy the usual group axioms. The theory $T$ is the smallest theory containing such $e, \nu, \mu$. A set $X$ is a group if there is a morphism of theories $T \to End(X)$. This means we have realizations of $e, \nu$, and $\mu$ on $X$.

DEFINITION 6.20. Let $X$ and $Y$ be $T$-algebras. Then a functor $H : X \to Y$ is a *morphism of $T$-algebras* in the categorical description if

$$\begin{CD}
Mor_T(m,n) @>>> Mor_{End(X)}(m,n) \\
@VVV @VV{H^{\times n}\circ}V \\
Mor_{End(Y)}(m,n) @>>{\circ H^{\times m}}> Functors(X^m, Y^n)
\end{CD}$$

commutes for all $m$ and $n$. A functor $H : X \to Y$ is a *morphism of $T$-algebras* in the functorial description if

$$\begin{CD}
T(m) @>>> End(X)(m) \\
@VVV @VV{H\circ}V \\
End(Y)(m) @>>{H^{\times m}}> Functors(X \times \cdots \times X, Y)
\end{CD}$$

commutes for all $m$.

EXAMPLE 6.21. Let $T$ be the theory of groups and let $X$ and $Y$ be groups. Then a set map $H : X \to Y$ is a morphism of $T$-algebras if and only if it is a group homomorphism.

THEOREM 6.22. *The category of categorical $T$-algebras is equivalent to the category of functorial $T$-algebras.*

*Proof:* The proof is similar to Theorem 6.17. □

Let $T$ be any theory. It is well known that $T$-algebras are algebras for a monad $C$, which depends on $T$. See for example [**39**] or [**44**]. We now present a version of this in preparation for the 2-monad whose strict algebras are pseudo $T$-algebras. Let $Cat_0$ denote the 1-category of small categories. We define a functor $C : Cat_0 \to Cat_0$ as follows. For a small category $X$, set

$$Obj\ CX := \frac{(\bigcup_{n\geq 0}(T(n) \times Obj\ X^n))}{\Gamma}$$

where the quotient by $\Gamma$ means to mod out by the smallest congruence satisfying $(w_f, x_1, \ldots, x_n) \sim (w, x_{f1}, \ldots, x_{fm})$ for all $m \in \mathbb{N}_0$, $w \in T(m)$, and maps $f : m = \{1, \ldots, m\} \to \{1, \ldots, n\} = n$. To define the morphisms of $CX$ we note that $\bigcup_{n\geq 0}(T(n) \times X^n)$ is a category if we interpret $T(n)$ as a discrete category for each $n$. Consider the directed graph with objects $Obj\ CX$ and arrows from $[a]$ to $[b]$ given by the union

$$\bigcup Mor_{\bigcup_{n\geq 0}(T(n)\times X^n)}(a', b')$$

over all $a' \sim a$ and $b' \sim b$. Next we take the free category on this directed graph and mod out by the relations of $\bigcup_{n\geq 0}(T(n) \times X^n)$ and the relations

$$(i_{w_f}, g_1, \ldots, g_n) = (i_w, g_{f1}, \ldots, g_{fm}).$$

This quotient category is $CX$. We define $C$ on functors $X \to Y$ analogously. Then $C : Cat_0 \to Cat_0$ is a functor because each step in the construction is functorial.

Next we define a natural transformation $\eta : 1_{Cat_0} \Rightarrow C$ by $\eta_X(x) := [1, x]$ for $x \in Obj\ X$ and $\eta_X(g) := [i_1, g]$ for a morphism $g$ in $X$. We also define a natural transformation $\mu : C^2 \Rightarrow C$ by

$$\mu_X([w, [v^1, x_1^1, \ldots, x_{j_1}^1], [v^2, x_1^2, \ldots, x_{j_2}^2], \ldots, [v^k, x_1^k, \ldots, x_{j_k}^k]]) :=$$

$$[\gamma(w, v^1, v^2, \ldots, v^k), x_1^1, \ldots, x_{j_k}^k]$$

for $w \in T(k), v^i \in T(j_i)$, and $(x_1^i, \ldots, x_{j_i}^i) \in X^{j_i}$ for $i = 1, \ldots, k$. On morphisms we define it to be

$$\mu_X([i_w, [i_{v^1}, g_1^1, \ldots, g_{j_1}^1], [i_{v^2}, g_1^2, \ldots, g_{j_2}^2], \ldots, [i_{v^k}, g_1^k, \ldots, g_{j_k}^k]]) :=$$

$$[i_{\gamma(w,v^1,v^2,\ldots,v^k)}, g_1^1, \ldots, g_{j_k}^k].$$

These assignments make $\mu_X : C^2X \to CX$ into a well defined functor because of the equivariances of $\gamma$. These natural transformations commute appropriately to make $C$ into a *monad* on the category $Cat_0$.

THEOREM 6.23. *The category of $C$-algebras is equivalent to the category of $T$-algebras.*

*Proof:* Let $\mathcal{C}_C$ and $\mathcal{C}_T$ denote the categories of $C$-algebras and $T$-algebras respectively. We construct a functor $\phi : \mathcal{C}_T \to \mathcal{C}_C$. Let $(X, \Phi)$ be a $T$-algebra. Then $\Phi_n : T(n) \to Functors(X^n, X)$ is a sequence of maps that is natural in $n$, preserves identity $1 \in T(1)$, and preserves compositions $\gamma$. This sequence of maps completely describes the algebraic structure. Let $h'$ denote the element of

$Functors(\bigcup_{n\geq 0}(T(n)\times X^n), X)$ that corresponds to the sequence under the bijection

$$\text{(6.1)} \qquad Functors(\bigcup_{n\geq 0}(T(n)\times X^n), X) \leftrightarrow \prod_{n\geq 0} Functors(T(n), X^{X^n}).$$

Then

$$h'(w_f, x_1, \ldots, x_n) = h'(w, x_{f1}, \ldots, x_{fm})$$
$$h'(i_{w_f}, g_1, \ldots, g_n) = h'(i_w, g_{f1}, \ldots, g_{fm})$$

because

$$\begin{array}{ccc} T(m) & \xrightarrow{\Phi_m} & Functors(X^m, X) \\ \Big\downarrow{\scriptstyle ()_f} & & \Big\downarrow{\scriptstyle ()_f} \\ T(n) & \xrightarrow[\Phi_n]{} & Functors(X^n, X) \end{array}$$

commutes. Hence $h' : \bigcup_{n\geq 0}(T(n)\times X^n) \to X$ induces a functor $h : CX \to X$, namely

$$[w, x_1, \ldots, x_m] \mapsto \Phi_m(w)(x_1, \ldots x_m)$$
$$[i_w, g_1, \ldots, g_m] \mapsto \Phi_m(i_w)(y_1, \ldots, y_m) \circ \Phi_m(w)(g_1, \ldots, g_m)$$
$$= \Phi_m(w)(g_1, \ldots, g_m)$$

for $g_i : x_i \to y_i$. Then $h : CX \to X$ makes $X$ into a $C$-algebra because the diagrams

$$\begin{array}{ccc} C^2X & \xrightarrow{Ch} & CX \\ \Big\downarrow{\scriptstyle \mu_X} & & \Big\downarrow{\scriptstyle h} \\ CX & \xrightarrow[h]{} & X \end{array} \qquad \begin{array}{ccc} X & \xrightarrow{\eta_X} & CX \\ & {\scriptstyle 1_X}\searrow & \Big\downarrow{\scriptstyle h} \\ & & X \end{array}$$

commute.

We define $\phi((X, \Phi)) := (X, h)$. For a morphism $H : (X, \Phi) \to (Y, \Psi)$ of $T$-algebras, let $\phi(H) : X \to Y$ be the same functor as $H$ on the underlying categories. Then

$$\begin{array}{ccc} CX & \xrightarrow{h_X} & X \\ \Big\downarrow{\scriptstyle C\phi(H)} & & \Big\downarrow{\scriptstyle \phi(H)} \\ CY & \xrightarrow[h_Y]{} & Y \end{array}$$

commutes. Then $\phi : \mathcal{C}_T \to \mathcal{C}_C$ is obviously a functor.

An "inverse" to $\phi$ can easily be constructed using the bijection (6.1). For example, let $(X, h)$ be a $C$-algebra. Then $h : CX \to X$ corresponds uniquely to a functor $h' : \bigcup_{n\geq 0}(T(n)\times X^n) \to X$ which satisfies

$$h'(w_f, x_1, \ldots, x_n) = h'(w, x_{f1}, \ldots, x_{fm})$$
$$h'(i_{w_f}, g_1, \ldots, g_n) = h'(w, g_{f1}, \ldots, g_{fm})$$

and $h'$ corresponds uniquely to some sequence $\Phi_n$ natural in $n$ which preserves 1 and $\gamma$.

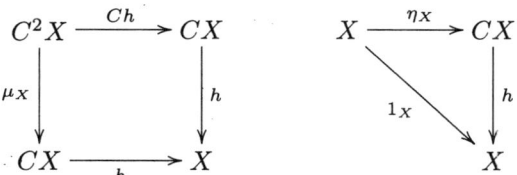

The equivalence of Theorem 6.22 yields the desired result. □

The concept of theory can be generalized to handle algebraic structures on more than one set, such as modules.

DEFINITION 6.24. A *theory on a set of objects* $J$, also called a *many-sorted theory*, is a category $\mathbf{T}$ whose objects are finite sequences $(j_1^{m_1}, \ldots, j_p^{m_p})$ with $j_1, \ldots, j_p \in J, p \geq 1$, and $m_1, \ldots, m_p \in \mathbb{N}_0$ such that $(j_1^{m_1}, \ldots, j_p^{m_p})$ is a product of copies of $j \in J$ where each $j$ appears $\sum_{r:j_r=j} m_r$ times. Each sequence is equipped with a limiting cone. Objects are equal to their reduced form, *e.g.* $(j^{m_1}, j^{m_2}) = (j^{m_1+m_2})$. We also abbreviate $(j^1) = j$.

EXAMPLE 6.25. An ordinary theory is a theory on one object, *i.e.* on the set $\{1\}$. We previously used $n$ to denote $1^n$ in the new notation.

EXAMPLE 6.26. Let $X_1$ and $X_2$ be categories. Then the *endomorphism theory* $End(X_j : j \in J)$ *on* $X_1$ *and* $X_2$ is an example of a theory on the set $J = \{1, 2\}$. The morphisms are

$$Mor_{End(X_j:j\in J)}((j_1^{m_1}, \ldots, j_p^{m_p}), (k_1^{n_1}, \ldots, k_q^{n_q})) :=$$

$$Functors(X_{j_1}^{m_1} \times \cdots \times X_{j_p}^{m^p}, X_{k_1}^{n_1} \times \cdots \times X_{k_q}^{n^q})$$

for $j_r, k_s \in \{1, 2\}$ and $m_r, n_s \in \mathbb{N}_0$. We easily see that $1^0$ and $2^0$ as well as $(1^0, 2^0)$ and $(2^0, 1^0)$ are terminal objects and that $(j_1^{m_1}, \ldots, j_p^{m_p})$ is a product of $\sum_{r:j_r=1} m_r$ copies of 1 and $\sum_{r:j_r=2} m_r$ copies of 2 equipped with the usual projections. Note also that there is a bijective correspondence.

$$Mor_{End(X_j:j\in J)}((j_1^{m_1}, \ldots, j_p^{m_p}), (k_1^{n_1}, \ldots, k_q^{n_q}))$$

$$\updownarrow$$

$$\prod_{r:k_r=1} Mor_{End(X_j:j\in J)}((j_1^{m_1}, \ldots, j_p^{m_p}), 1)^{\times n_r} \times \prod_{s:k_s=2} Mor_{End(X_j:j\in J)}((j_1^{m_1}, \ldots, j_p^{m_p}), 2)^{\times n_s}$$

In other words, the theory is determined by the sets

$$Mor_{End(X_j:j\in J)}((j_1^{m_1}, \ldots, j_p^{m_p}), 1) =: End(X_j : j \in J)_1(j_1^{m_1}, \ldots, j_p^{m_p})$$

$$Mor_{End(X_j:j\in J)}((j_1^{m_1}, \ldots, j_p^{m_p}), 2) =: End(X_j : j \in J)_2(j_1^{m_1}, \ldots, j_p^{m_p})$$

where $j_1, \ldots, j_p \in \{1, 2\}$ and $m_1, \ldots, m_p \in \mathbb{N}_0$ such that $j_r \neq j_{r+1}$ for all $1 \leq r \leq p - 1$.

Note also that for $n_1, \ldots, n_q \in \mathbf{N}_0$ and $k_1, \ldots, k_q \in J$ and maps

$$f : \sum_{r:j_r=1} m_r \to \sum_{r:k_r=1} n_r$$

$$g : \sum_{s:j_s=2} m_s \to \sum_{s:k_s=2} n_s$$

in $\Gamma$ we have *substitution* maps

$$End(X_j : j \in J)_1(j_1^{m_1}, \ldots, j_p^{m_p}) \xrightarrow{()_{f,g}} End(X_j : j \in J)_1(k_1^{n_1}, \ldots, k_q^{n_q})$$

$$End(X_j : j \in J)_2(j_1^{m_1}, \ldots, j_p^{m_p}) \xrightarrow{()_{f,g}} End(X_j : j \in J)_2(k_1^{n_1}, \ldots, k_q^{n_q}).$$

For example, let $w \in End(X_j : j \in J)_1(1^2, 2^2, 1^1, 2^2)$ and

$$f := \begin{pmatrix} 1 & 2 & 3 \\ 1 & 1 & 1 \end{pmatrix}, \quad g := \begin{pmatrix} 1 & 2 & 3 & 4 \\ 1 & 2 & 2 & 1 \end{pmatrix},$$

where $(k_1^{n_1}, k_2^{n_2}) = (1^1, 2^2)$ so that

$$f : 3 \to 1, \quad g : 4 \to 2.$$

Then $w_{f,g} \in End(X_j : j \in J)_1(1^1, 2^2)$ is defined by

$$\begin{aligned} w_{f,g}(x_1^1, x_1^2, x_2^2) &:= w(x_{f1}^1, x_{f2}^1, x_{g1}^2, x_{g2}^2, x_{f3}^1, x_{g3}^2, x_{g4}^2) \\ &= w(x_1^1, x_1^1, x_1^2, x_2^2, x_1^1, x_2^2, x_1^2). \end{aligned}$$

The notation $()_{f,g}$ suppresses the dependence of the map $()_{f,g}$ on $(j_1^{m_1}, \dots, j_p^{m_p})$ and $(k_1^{n_1}, \dots, k_q^{n_q})$.

There are also two *compositions* $\gamma_1$ and $\gamma_2$. For example

$$\gamma_1 : End(X_j : j \in J)_1(1^2, 2^2) \times End(X_j : j \in J)_1(\bar{n}_1) \times End(X_j : j \in J)_1(\bar{n}_2) \times$$
$$\times End(X_j : j \in J)_2(\bar{n}_3) \times End(X_j : j \in J)_2(\bar{n}_4) \to End(X_j : j \in J)_1(\bar{n}_1 \cdot \bar{n}_2 \cdot \bar{n}_3 \cdot \bar{n}_4)$$

and

$$\gamma_2 : End(X_j : j \in J)_2(2^3, 1^1) \times End(X_j : j \in J)_2(\bar{n}_1) \times End(X_j : j \in J)_2(\bar{n}_2) \times$$
$$\times End(X_j : j \in J)_2(\bar{n}_3) \times End(X_j : j \in J)_1(\bar{n}_4) \to End(X_j : j \in J)_2(\bar{n}_1 \cdot \bar{n}_2 \cdot \bar{n}_3 \cdot \bar{n}_4)$$

where $\bar{n}_1 \cdot \bar{n}_2 \cdot \bar{n}_3 \cdot \bar{n}_4$ means to concatenate the objects $\bar{n}_1, \dots, \bar{n}_4$ and to reduce, *e.g.* $(1^1, 2^2) \cdot (2^3, 1^2) = (1^1, 2^5, 1^2)$.

There are also units $1_1 \in End(X_j : j \in J)_1(1)$ and $1_2 \in End(X_j : j \in J)_2(2)$.

The compositions are associative, unital, and equivariant. The substitution is also functorial. This example easily extends to arbitrary $J$.

DEFINITION 6.27. Let $\Gamma_J$ denote the category whose objects are finite sequences $(j_1^{m_1}, \dots, j_p^{m_p})$ with $j_1, \dots, j_p \in J, p \geq 1$, and $m_1, \dots, m_p \in \mathbb{N}_0$. Objects are equal to their reduced form, *e.g.* $(j^{m_1}, j^{m_2}) = (j^{m_1+m_2})$. We also abbreviate $(j^1) = j$. The morphisms are

$$Mor_{\Gamma_J}((j_1^{m_1}, \dots, j_p^{m_p}), (k_1^{n_1}, \dots, k_q^{n_q})) := \prod_{\ell \in J} Mor_\Gamma\Big(\sum_{r : j_r = \ell} m_r, \sum_{s : k_s = \ell} n_s\Big)$$

where $\Gamma$ denotes the category in Definition 6.7.

In this definition the hom sets are assumed to be disjoint.

Several of the results on theories carry over to these generalized theories on a set of objects.

THEOREM 6.28. *A theory* $\mathbf{T}$ *on a set of objects* $J$ *is equivalent to a collection of functors* $\{\mathbf{T}_j : \Gamma_J \to Sets | j \in J\}$ *equipped with compositions*

$$\gamma_j : \mathbf{T}_j(j_1^{k_1}, \dots, j_p^{k_p}) \times \mathbf{T}_{j_1}(\bar{n}_1^1) \times \dots \times \mathbf{T}_{j_1}(\bar{n}_{k_1}^1) \times$$
$$\times \mathbf{T}_{j_2}(\bar{n}_1^2) \times \dots \times \mathbf{T}_{j_2}(\bar{n}_{k_2}^2) \times$$
$$\dots$$
$$\times \mathbf{T}_{j_p}(\bar{n}_1^p) \times \dots \times \mathbf{T}_{j_p}(\bar{n}_{k_p}^p) \to \mathbf{T}_j(\bar{n}_1^1 \cdots \bar{n}_{k_1}^1 \cdot \bar{n}_1^2 \cdots \bar{n}_{k_2}^2 \cdots \bar{n}_1^p \cdots \bar{n}_{k_p}^p)$$

*for each* $j \in J$ *and* $(j_1^{k_1}, \dots, j_p^{k_p}), \bar{n}_1^1, \dots, \bar{n}_{k_p}^p \in Obj\, \Gamma_J$ *and equipped with units* $1_j \in \mathbf{T}_j(j)$ *for each* $j \in J$ *which satisfy analogues of (1) through (5) in Lemma 6.6. Elements of* $\mathbf{T}_j(\bar{n})$ *are called words.*

*Proof:* Set $\mathbf{T}_j(\bar{n}) := Mor_{\mathbf{T}}(\bar{n}, j)$ and proceed like in the case of a theory on the set $\{1\}$. $\square$

EXAMPLE 6.29. The *theory* $\mathbf{R}$ *of theories* is a theory on the set $\mathbb{N}_0$. There are three types of generating morphisms.

- For each $k \geq 1$ and $n_1, \ldots, n_k \geq 0$ there is a morphism $\gamma : (k, n_1, \ldots, n_k) \to (n_1 + \cdots + n_k)$ called *composition*.
- For each $f : m \to n$ in $\Gamma$ there is a morphism $()_f : (m) \to (n)$ called *substitution*.
- There is a morphism $1 : (1^0) \to (1^1)$ called the *unit*.

The substitution and unit are not to be confused with the substitution and units with which every theory on a set of objects is equipped. These morphisms must satisfy the relations of theories in Lemma 6.6, namely associativity, equivariances, unitality, and functoriality.

Next we can speak of morphisms of theories on the set $J$ as well as algebras for theories on the set $J$ just as in the case $J = \{1\}$.

DEFINITION 6.30. A *morphism of theories on a set* $J$ is a functor $\Phi : \mathbf{S} \to \mathbf{T}$ such that $\Phi(j_1^{m_1}, \ldots, j_p^{m_p}) = (j_1^{m_1}, \ldots, j_p^{m_p})$ and $\Phi(pr) = pr$ for every projection.

THEOREM 6.31. *The analogue of Theorem 6.17 holds for theories on a set of objects $J$.*

DEFINITION 6.32. Let $\mathbf{T}$ be a theory on the set $J$ and $\{X_j | j \in J\}$ a collection of categories. Then $\{X_j\}_j$ form an *algebra over* $\mathbf{T}$ or a $\mathbf{T}$*-algebra* if they are equipped with a morphism $\Phi : \mathbf{T} \to End(X_j : j \in J)$ of theories on $J$.

EXAMPLE 6.33. Let $\mathbf{R}$ denote the theory of theories. Let $T$ be a theory. Then $\{T(j) | j \in \mathbb{N}_0\}$ form an $\mathbf{R}$-algebra. In other words, a theory is an algebra over the theory of theories. A morphism of theories is nothing more than a morphism of algebras over the theory of theories.

THEOREM 6.34. *The analogue of Theorem 6.22 holds for a theory* $\mathbf{T}$ *on a set of objects.*

We can use the theory $\mathbf{R}$ of theories to construct a monad $C$ on the category $\prod_{n \geq 0} Sets$ whose algebras are the usual theories. In fact, $CT$ is the sequence of sets underlying the *free theory* on $T$. This free theory is essential to several of the proofs in this paper. Let $T = (T(n))_{n \geq 0}$ be an object of $\prod_{n \geq 0} Sets$ and $J := \mathbb{N}_0$. Then the *free theory* on $T$ is defined by

$$CT(n) := \frac{\bigcup_{\bar{m} \in Obj\ \Gamma_J} \mathbf{R}_n(\bar{m}) \times T(j_1)^{\times m_1} \times \cdots \times T(j_p)^{\times m_p}}{\Gamma_J}$$

where $\bar{m} = (j_1^{m_1}, \ldots, j_p^{m_p})$.

We can generalize the notion of theory in yet another direction. Instead of considering arbitrary sets $J$, we can consider theories which are also 2-categories in which every 2-cell is iso. We will use these to describe pseudo algebras in a compact way. See [**44**] for a more general concept of enriched Lawvere theory.

DEFINITION 6.35. A *theory enriched in groupoids* is a 2-category $\mathcal{T}$ with iso 2-cells and with objects $0, 1, 2, \ldots$ such that $n$ is the 2-product of 1 with itself $n$ times in the 2-category $\mathcal{T}$ and each $n$ is equipped with a limiting 2-cone.

This definition means for each $n \in Obj\,\mathcal{T}$ we have chosen morphisms $\pi_i^n = pr_i : n \to 1$ for $i = 1, \ldots, n$ with the universal property that

$$Mor_{\mathcal{T}}(m,n) \xrightarrow{\pi^n \circ} 2 - Cone(m,F)$$

is an isomorphism for all $m \in Obj\,\mathcal{T}$, where $F : \{1,\ldots,n\} \to \mathcal{T}$ is the 2-functor which is constant 1. It is tempting to call such a theory a 2-theory, but we reserve that name for something else. As before, we use the notation $\mathcal{T}(n)$ for the category $Mor_{\mathcal{T}}(n,1)$. Using the universal property, we can construct $\prod$ and $(\ldots)$ for the 2-cells. For any object $m \in Obj\,T$, morphisms $w_i, v_i : m \to 1$, and 2-cells $\alpha_i : w_i \Rightarrow v_i$ for $i = 1, \ldots, n$, there exists a unique 2-cell $\prod_{j=1}^n \alpha_j : \prod_{j=1}^n w_j \Rightarrow \prod_{j=1}^n v_j$ such that

$$i_{pr_i} * \prod_{j=1}^{n} \alpha_j = \alpha_i$$

for all $i = 1, \ldots n$. For any $k \in \mathbb{N}_0$, any morphisms $w_i, v_i : n_i \to 1$, and any 2-cells $\alpha_i : w_i \Rightarrow v_i$ for $i = 1, \ldots, k$, there is a unique 2-cell $(\alpha_1, \ldots, \alpha_k) : (w_1, \ldots, w_k) \Rightarrow (v_1, \ldots, v_k)$ such that

$$i_{pr_i} * (\alpha_1, \ldots, \alpha_k) = (\alpha_i)_{\iota_i}$$

for all $i = 1, \ldots, k$.

EXAMPLE 6.36. Let $X$ be a category. Then the *endomorphism theory $End(X)$ enriched in groupoids* has objects $0, 1, 2, \ldots$, morphisms $Obj\, Mor_{End(X)}(m,n) = Functors(X^m, X^n)$ and 2-cells the natural isomorphisms.

Most of the work on theories carries over to the enriched context with minor additions for the 2-cells. The statements of the relevant theorems are as follows. The term *map* is simply replaced by *functor*.

LEMMA 6.37. *Let $\mathcal{T}$ be a theory enriched in groupoids. Then the morphism category $Mor_{\mathcal{T}}(m,n)$ is isomorphic to the product category $\prod_{j=1}^n Mor_{\mathcal{T}}(m,1)$.*

LEMMA 6.38. *Let $\mathcal{T}$ be a theory enriched in groupoids. Then for all $k, n_1, \ldots, n_k \in \{0,1,\ldots\}$ there is a functor $\gamma : \mathcal{T}(k) \times \mathcal{T}(n_1) \times \cdots \times \mathcal{T}(n_k) \to \mathcal{T}(n_1 + \cdots + n_k)$ called composition and for every function $f : \{1,\ldots,k\} \to \{1,\ldots,\ell\}$ there is a functor $\mathcal{T}(k) \xrightarrow{()_f} \mathcal{T}(\ell)$ called substitution. These functors satisfy the enriched analogues of (1) through (5) in Lemma 6.6.*

*Proof:* Define $\gamma(w, w_1, \ldots, w_k) := w \circ (w_1, \ldots, w_k)$ as before. Additionally, define $\gamma(\alpha, \alpha_1, \ldots, \alpha_k) := \alpha * (\alpha_1, \ldots, \alpha_k)$ for 2-cells. Define $w_f := w \circ f'$ as before and $\alpha_f := \alpha * i_{f'}$ where $i_{f'} : f' \Rightarrow f'$ is the identity 2-cell of the morphism $f'$ in $\mathcal{T}$ and $\alpha : w \Rightarrow v$ is a 2-cell. The rest of proof is similar to Lemma 6.6. □

LEMMA 6.39. *Let $\mathcal{T}$ be a 2-functor from $\Gamma$ to the 2-category $Cat$ of small categories equipped with functors $\gamma : \mathcal{T}(k) \times \mathcal{T}(n_1) \times \cdots \times \mathcal{T}(n_k) \to \mathcal{T}(n_1 + \cdots + n_k)$ and an object $1 \in \mathcal{T}(1)$ which satisfy (1) through (5) of Lemma 6.6 where $()_f := \mathcal{T}(f)$ for functions $f : k \to \ell$. Then $\mathcal{T}$ determines a theory enriched in groupoids with $Mor(n,1) = \mathcal{T}(n)$ for all $n \geq 0$.*

THEOREM 6.40. *A theory $\mathcal{T}$ enriched in groupoids is determined by either of the following equivalent collections of data:*

(1) *A 2-category $\mathcal{T}$ with objects $0, 1, 2, \ldots$ such that $n$ is the 2-categorical product of $1$ with itself $n$ times and each $n$ is equipped with a limiting 2-cone.*
(2) *A 2-functor $\mathcal{T} : \Gamma \to Cat$ equipped with functors $\gamma : \mathcal{T}(k) \times \mathcal{T}(n_1) \times \cdots \times \mathcal{T}(n_k) \to \mathcal{T}(n_1 + \cdots + n_k)$ and a unit $1 \in \mathcal{T}(1)$ which satisfy (1) through (5) of Lemma 6.6.*

*Proof:* In each description $Mor_{\mathcal{T}}(n, 1)$ is the same. By the universality of 2-products this determines the rest of the theory. □

DEFINITION 6.41. Let $\mathcal{S}$ and $\mathcal{T}$ be theories enriched in groupoids. In the 2-categorical description of $\mathcal{S}$ and $\mathcal{T}$ a *morphism of theories enriched in groupoids* $\Phi : \mathcal{S} \to \mathcal{T}$ is a 2-functor from the 2-category $\mathcal{S}$ to the 2-category $\mathcal{T}$ such that $\Phi(n_{\mathcal{S}}) = n_{\mathcal{T}}$ and $\Phi(pr_i) = pr_i$ for all projections.

The analogue for Lemma 6.13 incorporates the 2-cells below.

LEMMA 6.42. *Let $\Phi : \mathcal{S} \to \mathcal{T}$ be a morphism of theories enriched in groupoids.*
(1) *Let $f : \{1, \ldots, k\} \to \{1, \ldots, \ell\}$ be a function. As usual, $f' : \ell \to k$ denotes the unique morphism in any theory such that*

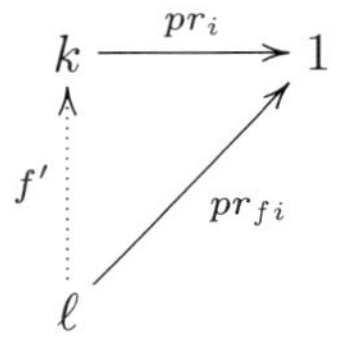


*commutes. Then $\Phi(f') = f'$.*
(2) *Let $f : \{1, \ldots, k\} \to \{1, \ldots, \ell\}$ be a function and $w \in Mor_{\mathcal{S}}(k, 1)$. Then $\Phi(w_f) = \Phi(w)_f$.*
(3) *Let $w_j, v_j \in Mor_{\mathcal{S}}(m, 1)$ and $\alpha_j : w_j \Rightarrow v_j$ for $j = 1, \ldots, n$. Then $\Phi(\prod_{j=1}^n w_j) = \prod_{j=1}^n \Phi(w_j)$ and $\Phi(\prod_{j=1}^n \alpha_j) = \prod_{j=1}^n \Phi(\alpha_j)$.*
(4) *Let $w_j, v_j \in Mor_{\mathcal{S}}(n_j, 1)$ for $j = 1, \ldots, k$. Then we have $\Phi(w_1, \ldots, w_k) = (\Phi(w_1), \ldots, \Phi(w_k))$ and $\Phi(\alpha_1, \ldots, \alpha_k) = (\Phi(\alpha_1), \ldots, \Phi(\alpha_k))$.*

THEOREM 6.43. *Let $\mathcal{S}$ and $\mathcal{T}$ be theories enriched in groupoids. Then a morphism $\mathcal{S} \to \mathcal{T}$ of theories enriched in groupoids is given by either of the following equivalent collections of data:*
(1) *A 2-functor $\Phi : \mathcal{S} \to \mathcal{T}$ such that $\Phi(n_{\mathcal{S}}) = n_{\mathcal{T}}$ for all $n_{\mathcal{S}} \in Obj\ \mathcal{S}$ and $\Phi(pr_i) = pr_i$ for all projections*
(2) *A 2-natural transformation $\Phi : \mathcal{S} \Rightarrow \mathcal{T}$ of the 2-functors $\mathcal{S}, \mathcal{T} : \Gamma \to Cat$ which preserves the $\gamma$'s and the units.*

THEOREM 6.44. *The 2-category of theories enriched in groupoids with objects and morphisms as in (1) of Theorems 6.40 and 6.43 is 2-equivalent to the 2-category with objects and morphisms as in (2) of Theorems 6.40 and 6.43.*

We can now define algebras over theories enriched in groupoids in analogy to algebras over theories.

DEFINITION 6.45. Let $X$ be a category and $\mathcal{T}$ a theory enriched over groupoids. Then $X$ is a *$\mathcal{T}$-algebra* if it is equipped with a morphism of theories $\mathcal{T} \to End(X)$ enriched in groupoids. We also say $X$ is an *algebra over the theory $\mathcal{T}$*.

Our main example, pseudo $T$-algebras, will be given in the next chapter as strict $\mathcal{T}$-algebras, where $\mathcal{T}$ is obtained from the free theory on $T$.

THEOREM 6.46. *The analogue of Theorem 6.22 holds for theories enriched in groupoids.*

CHAPTER 7

# Pseudo $T$-Algebras

In this chapter we introduce the 2-category of pseudo $T$-algebras for a theory $T$. A *pseudo algebra* in this paper is the same thing as a *lax algebra* in [**25**], [**26**], and [**27**]. We construct from $T$ a theory $\mathcal{T}$ enriched in groupoids and show that a pseudo algebra over $T$ is the same thing as an algebra over $\mathcal{T}$. Theorem 7.14 says that the 2-category of pseudo $T$-algebras and pseudo morphisms is 2-equivalent to the 2-category of strict $C$-algebras with pseudo morphisms for the 2-monad $C$ defined on page 70. This 2-category of strict $C$-algebras and pseudo morphisms admits pseudo limits by a result of Blackwell, Kelly, and Power in [**9**]. Hence the 2-category of pseudo $T$-algebras admits pseudo limits. In the next chapter we give a concrete construction of a pseudo limit. For more on pseudo algebras over 2-monads see [**24**], [**32**], and [**33**].

DEFINITION 7.1. Let $T$ be a theory. A category $X$ is a *pseudo $T$-algebra* or a *pseudo algebra over $T$* if it is equipped with *structure maps* $\Phi_n : T(n) \longrightarrow Functors(X^n, X)$ for every $n \in \mathbb{N}$ as well as the coherence isomorphisms below. Moreover, the coherence isomorphisms are required to satisfy the coherence diagrams below. We write simply $\Phi$ for all $\Phi_n$. The coherence isomorphisms are indexed by the operations of theories and are as follows:

(1) For every $k \in \mathbb{N}$, $w \in T(k)$, and all words $w_1, \ldots, w_k$, there is a natural isomorphism $c_{w,w_1,\ldots,w_k} : \Phi(\gamma(w, w_1, \ldots, w_k)) \Rightarrow \gamma(\Phi(w), \Phi(w_1), \ldots, \Phi(w_k))$. This means that $\Phi$ preserves composition up to a natural isomorphism.
(2) There is a natural isomorphism $I : \Phi(1) \Rightarrow 1_X$ where 1 is the identity word and $1_X$ is the identity functor $X \to X$. This means that $\Phi$ preserves the identity up to a natural isomorphism.
(3) For every word $w \in T(m)$ and function $f : \{1, \ldots, m\} \to \{1, \ldots, n\}$, there is a natural isomorphism $s_{w,f} : \Phi(w_f) \Rightarrow \Phi(w)_f$ where the substituted functor $\Phi(w)_f : X^n \to X$ is defined in Examples and 6.3 and 6.4. This means that $\Phi$ preserves the substitution up to a natural isomorphism.

The coherence diagrams are indexed by relations of theories and are as follows. The commutivity of these diagrams means that they commute when evaluated on every tuple of objects of $X$ of appropriate length.

(1) The composition coherence isomorphisms are associative. For example, for $u, v, w \in T(1)$ the diagram below must commute where $i_F$ means the identity natural transformation $F \to F$ for a functor $F$.

$$\begin{array}{ccc}
\Phi(\gamma(w,\gamma(v,u))) = \Phi(\gamma(\gamma(w,v),u)) & \xRightarrow{c_{\gamma(w,v),u}} & \gamma(\Phi(\gamma(w,v)),\Phi(u)) \\
\Big\Downarrow {\scriptstyle c_{w,\gamma(v,u)}} & & \Big\Downarrow {\scriptstyle \gamma(c_{w,v},i_{\Phi(u)})} \\
\gamma(\Phi(w),\Phi(\gamma(v,u))) \xRightarrow[\gamma(i_{\Phi(w)},c_{v,u})]{} & \gamma(\Phi(w),\gamma(\Phi(v),\Phi(u))) & = \gamma(\gamma(\Phi(w),\Phi(v)),\Phi(u))
\end{array}$$

(2) The natural isomorphism for the identity word commutes with the natural isomorphism for the composition, *i.e.* for every $n \in \mathbb{N}$ and every word $w \in T(n)$ the diagram below must commute where $1_X$ is the identity functor on $X$.

$$\begin{array}{ccc}
\Phi(\gamma(w,1,\ldots,1)) & =\!=\!=\!=\!= & \Phi(w) \\
\Big\Downarrow {\scriptstyle c_{w,1,\ldots,1}} & & \Big\| \\
\gamma(\Phi(w),\Phi(1),\ldots,\Phi(1)) & \xRightarrow[\gamma(i_{\Phi(w)},I,\ldots,I)]{} & \gamma(\Phi(w),1_X,\ldots,1_X)
\end{array}$$

(3) The natural isomorphism for the identity word commutes with the natural isomorphism for the composition also in the sense that for every word $w \in T(n)$ the diagram below must commute.

$$\begin{array}{ccc}
\Phi(\gamma(1,w)) & =\!=\!=\!=\!= & \Phi(w) \\
\Big\Downarrow {\scriptstyle c_{1,w}} & & \Big\| \\
\gamma(\Phi(1),\Phi(w)) & \xRightarrow[\gamma(I,i_{\Phi(w)})]{} & \gamma(1_X,\Phi(w))
\end{array}$$

(4) Let $f : \{1,\ldots,k\} \to \{1,\ldots,\ell\}$ be a function and let $\bar{f} : \{1,2,\ldots,n_{f1}+n_{f2}+\cdots+n_{fk}\} \to \{1,2,\ldots,n_1+n_2+\cdots+n_\ell\}$ be the function that moves entire blocks according to $f$ as in Example 6.3. Then equivariance is preserved in the sense that the diagram below must commute.

$$\begin{array}{ccc}
\Phi(\gamma(w,w_{f1},\ldots,w_{fk})_{\bar{f}}) & \xRightarrow{s_{\gamma(w,w_{f1},\ldots,w_{fk}),\bar{f}}} & \Phi(\gamma(w,w_{f1},\ldots,w_{fk}))_{\bar{f}} \\
\Big\| & & \Big\Downarrow {\scriptstyle (c_{w,w_{f1},\ldots,w_{fk}})_{\bar{f}}} \\
\Phi(\gamma(w_f,w_1,\ldots,w_\ell)) & & \gamma(\Phi(w),\Phi(w_{f1}),\Phi(w_{fk}))_{\bar{f}} \\
\Big\Downarrow {\scriptstyle c_{w_f,w_1,\ldots,w_\ell}} & & \Big\| \\
\gamma(\Phi(w_f),\Phi(w_1),\ldots,\Phi(w_\ell)) & \xRightarrow[\gamma(s_{w,f},i_{\Phi(w_1)},\ldots,i_{\Phi(w_\ell)})]{} & \gamma(\Phi(w)_f,\Phi(w_1),\ldots,\Phi(w_\ell))
\end{array}$$

(5) Let $g_i : \{1,\ldots,n_i\} \to \{1,\ldots,n_i'\}$ be functions and let $g_1+\cdots+g_k : \{1,2,\ldots,n_1+\cdots+n_k\} \to \{1,2,\ldots,n_1'+\cdots+n_k'\}$ be the function obtained by placing $g_1,\ldots,g_k$ next to each other from left to right. Then equivariance is preserved in the sense that the diagram below must commute.

$$
\begin{array}{ccc}
\Phi(\gamma(w,w_1,\ldots,w_k)_{g_1+\cdots+g_k}) & \xRightarrow{s_{\gamma(w,w_1,\ldots,w_k),g_1+\cdots+g_k}} & \Phi(\gamma(w,w_1,\ldots,w_k))_{g_1+\cdots+g_k} \\
\| & & \Downarrow (c_{w,w_1,\ldots,w_k})_{g_1+\cdots+g_k} \\
\Phi(\gamma(w,(w_1)_{g_1},\ldots,(w_k)_{g_k})) & & \gamma(\Phi(w),\Phi(w_1),\ldots,\Phi(w_k))_{g_1+\cdots+g_k} \\
\Downarrow c_{w,(w_1)_{g_1},\ldots,(w_k)_{g_k}} & & \| \\
\gamma(\Phi(w),\Phi((w_1)_{g_1}),\ldots,\Phi((w_k)_{g_k})) & \xRightarrow[\gamma(i_{\Phi(w)},s_{w_1,g_1},\ldots,s_{w_k,g_k})]{} & \gamma(\Phi(w),\Phi(w_1)_{g_1},\ldots,\Phi(w_k)_{g_k})
\end{array}
$$

(6) The substitution coherence isomorphisms are associative, *i.e.* for every word $w \in T(\ell)$ and functions $f : \{1,\ldots,\ell\} \to \{1,\ldots,m\}$ and $g : \{1,\ldots,m\} \to \{1,\ldots,n\}$ we mimic the equality $w_{g\circ f} = (w_f)_g$ by requiring the diagram below to commute. Here $(s_{w,f})_g$ is the natural transformation which is defined for objects $A_1,\ldots,A_n$ of $X$ by $(s_{w,f})_g(A_1,\ldots,A_n) = s_{w,f}(A_{g1},\ldots,A_{gm})$.

$$
\begin{array}{ccc}
\Phi((w_f)_g) = \Phi(w_{g\circ f}) & \xRightarrow{s_{w,g\circ f}} & \Phi(w)_{g\circ f} \\
\Downarrow s_{(w_f),g} & & \| \\
\Phi(w_f)_g & \xRightarrow[(s_{w,f})_g]{} & (\Phi(w)_f)_g
\end{array}
$$

(7) For all $w \in T(k)$ and $id_k : \{1,\ldots,k\} \to \{1,\ldots,k\}$ the natural transformation $s_{w,id_k}$ is the identity.

REMARK 7.2. One can compactly describe the concept of a pseudo algebra as follows. A category $X$ is a pseudo $T$-algebra if it is equipped with a *pseudo morphism of theories* $\Phi : T \to End(X)$. The assignment $\Phi$ is pseudo in the sense that the requirements of Lemma 6.14 are only satisfied up to coherence isos, namely the assignment preserves $\gamma$ up to $c$, preserves the identity up $I$, and is natural up to $s$ as in the diagrams below and these coherence isos satisfy coherence diagrams.

$$
\begin{array}{ccc}
T(k)\times T(n_1)\times\cdots\times T(n_k) & \xrightarrow{\Phi(k)\times\Phi(n_1)\times\cdots\times\Phi(n_k)} & End(X)(k)\times End(X)(n_1)\times\cdots\times End(X)(n_k) \\
\downarrow \gamma^T & \overset{c}{\nearrow\!\!\!\nearrow} & \downarrow \gamma^{End(X)} \\
T(n_1+\cdots+n_k) & \xrightarrow[\Phi(n_1+\cdots+n_k)]{} & End(X)(n_1+\cdots+n_k)
\end{array}
$$

$$
\begin{array}{ccc}
X & = & X \\
\downarrow \Phi(1)(1_T) & \overset{I}{\nearrow\!\!\!\nearrow} & \downarrow 1_X \\
X & = & X
\end{array}
$$

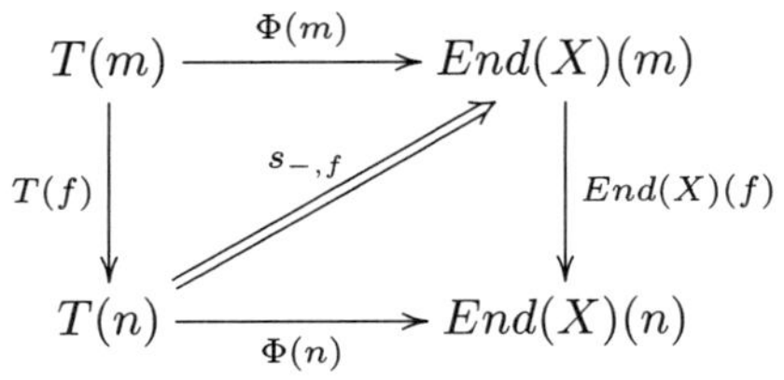


REMARK 7.3. It is possible to describe the general form of these coherence diagrams. In general, a relation $\alpha \circ \beta = \alpha' \circ \beta'$ in the theory of theories and a tuple $\bar{w}$ of words gives rise to a coherence diagram

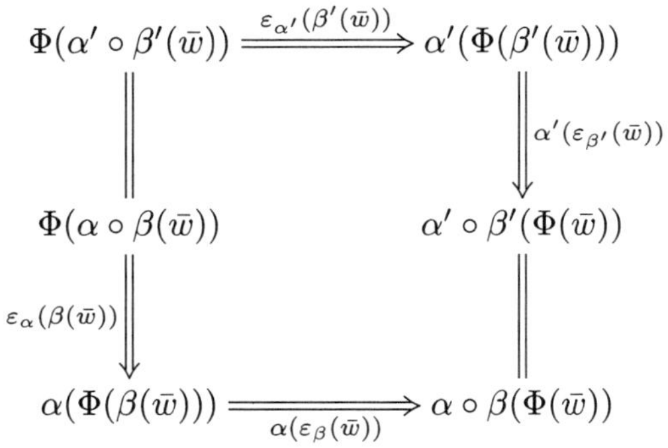


where $\varepsilon_\alpha, \varepsilon_{\alpha'}, \varepsilon_\beta$, and $\varepsilon_{\beta'}$ are the coherence isos associated to the morphisms $\alpha$, $\alpha'$, $\beta$, and $\beta'$ respectively in the theory of theories and $\Phi(\bar{w})$ denotes the tuple of words obtained by applying $\Phi$ to each of the constituents of $\bar{w}$. Note that $\varepsilon_\alpha, \varepsilon_{\alpha'}, \varepsilon_\beta$, and $\varepsilon_{\beta'}$ are tuples of the 2-cells $c, I, s$ and identity 2-cells. In the definition of pseudo algebra above, the morphisms $\beta, \beta'$ are tuples of generating morphisms in all cases except in (4). In (4) the $\beta'$ is the result of applying a substitution morphism in the theory of theories to $\gamma$. This substitution morphism can be written in terms of $f$ appropriately. In this case we have $\varepsilon_{\beta'}(\bar{w}) = c_{w,w_{f1},\ldots,w_{fk}}$.

DEFINITION 7.4. Let $X$ and $Y$ be pseudo $T$-algebras and $H : X \to Y$ a functor between the underlying categories. Denote the structure maps of $X$ and $Y$ by $\Phi$ and $\Psi$ respectively. For all $n \in \mathbb{N}$ and all $w \in T(n)$ let $\rho_w : H \circ \Phi(w) \Rightarrow \Psi(w) \circ (H, \ldots, H)$ be a natural isomorphism. Then $H$ is a *pseudo morphism of pseudo $T$-algebras with coherence iso 2-cells $\rho_w$* (or just *morphism of pseudo $T$-algebras* for short) if the following coherence diagrams of natural isomorphisms are satisfied.

(1) For all $k \in \mathbb{N}$, $w \in T(k)$, and all words $w_1, \ldots, w_k$ of $T$ the diagram below must commute.

$$\begin{array}{ccc}
H \circ \Phi(w \circ (w_1, \ldots, w_k)) & \xRightarrow{i_H * c_{w,w_1,\ldots,w_k}} & H \circ \Phi(w) \circ (\Phi(w_1), \ldots, \Phi(w_k)) \\
 & & \Big\Downarrow {\scriptstyle \rho_w * i_{(\Phi(w_1),\ldots,\Phi(w_k))}} \\
{\scriptstyle \rho_{w \circ (w_1,\ldots,w_k)}} \Big\Downarrow & & \Psi(w) \circ (H, \ldots, H) \circ (\Phi(w_1), \ldots, \Phi(w_k)) \\
 & & \Big\Downarrow {\scriptstyle i_{\Psi(w)} * (\rho_{w_1},\ldots,\rho_{w_k})} \\
\Psi(w \circ (w_1, \ldots, w_k)) \circ (H, \ldots, H) & \xRightarrow[c_{w,w_1,\ldots,w_k} * i_{(H,\ldots,H)}]{} & \Psi(w) \circ (\Psi(w_1), \ldots, \Psi(w_k)) \circ (H, \ldots, H)
\end{array}$$

(2) The diagram below must commute.

$$\begin{array}{ccc}
H \circ \Phi(1) & \xRightarrow{i_H * I} & H \circ 1_X \\
{\scriptstyle \rho_1} \Big\Downarrow & & \Big\| \\
\Psi(1) \circ H & \xRightarrow[I * i_H]{} & 1_Y \circ H
\end{array}$$

(3) For every word $w \in T(m)$ and every function $f : \{1, \ldots, m\} \to \{1, \ldots, n\}$ the diagram below must commute.

$$\begin{array}{ccc}
H \circ \Phi(w_f) & \xRightarrow{i_H * s_{w,f}} & H \circ \Phi(w)_f \\
{\scriptstyle \rho_{w_f}} \Big\Downarrow & & \Big\Downarrow {\scriptstyle (\rho_w)_f} \\
\Psi(w_f) \circ (H, \ldots, H) & \xRightarrow[s_{w,f} * i_{(H,\ldots,H)}]{} & \Psi(w)_f \circ (H, \ldots, H)
\end{array}$$

EXAMPLE 7.5. Let $T$ be the theory of commutative monoids and let $FiniteSets$ be the category of finite sets and bijections. Define $A \coprod B := A \times \{1\} \cup B \times \{2\}$ for finite sets $A$ and $B$. Define coproduct similarly for morphisms of finite sets. Then $\coprod : FiniteSets \times FiniteSets \to FiniteSets$ is a functor which makes $FiniteSets$ into a pseudo $T$-algebra, *i.e.* a *pseudo commutative monoid.* More generally, any symmetric monoidal category is a pseudo $T$-algebra.

EXAMPLE 7.6. Let $T$ be the theory of commutative semi-rings. Then the category of finite dimensional complex vector spaces is a pseudo $T$-algebra whose structure is given by direct sum and tensor product. We also say this category is a *pseudo commutative semi-ring.*

DEFINITION 7.7. Let $X, Y$, and $Z$ be pseudo $T$-algebras and $G : X \to Y, H : Y \to Z$ morphisms of pseudo $T$-algebras with coherence 2-cells $\rho_w^G$ and $\rho_w^H$ respectively. Then the *composition* $H \circ G$ is the composition of the underlying functors . It has the coherence 2-cells $\rho_w^{H \circ G} := (\rho_w^H * i_{(G,\ldots,G)}) \odot (i_H * \rho_w^G) : H \circ G \circ \Phi(w) \Rightarrow \Psi(w) \circ (H \circ G, \ldots, H \circ G)$ where $\Phi$ ad $\Psi$ denote the structure maps of $X$ and $Z$ respectively.

LEMMA 7.8. *The composition of morphisms of pseudo $T$-algebras is a morphism of pseudo $T$-algebras.*

*Proof:* Immediate. □

DEFINITION 7.9. Let $X$ and $Y$ be pseudo $T$-algebras with structure maps $\Phi$ and $\Psi$ respectively. Let $G, H : X \to Y$ be morphisms of pseudo $T$-algebras. A natural transformation $\alpha : G \Rightarrow H$ between the underlying functors is a *2-cell in the 2-category of pseudo $T$-algebras* if for all $n \in \mathbb{N}$ and all $w \in T(n)$

$$\begin{array}{ccc} G \circ \Phi(w) & \xRightarrow{\alpha * i_{\Phi(w)}} & H \circ \Phi(w) \\ \Big\Downarrow \rho_w^G & & \Big\Downarrow \rho_w^H \\ \Psi(w) \circ (G, \ldots, G) & \xRightarrow[i_{\Psi(w)} * (\alpha, \ldots, \alpha)]{} & \Psi(w) \circ (H, \ldots, H) \end{array}$$

commutes. The vertical and horizontal compositions of the 2-cells are just the vertical and horizontal composition of the underlying natural transformations.

LEMMA 7.10. *The small pseudo $T$-algebras with morphisms and 2-cells defined above form a 2-category.*

*Proof:* The axioms can be verified directly. □

Next we work towards a description of pseudo $T$-algebras as strict algebras over a *2-monad* $C$ by way of a theory $\mathcal{T}$ enriched in groupoids. As mentioned in the last chapter, a pseudo $T$-algebra is the same thing as a strict $\mathcal{T}$-algebra. This was observed in [**27**]. We can see this as follows. Let $T'$ denote the free theory on the sequence of sets underlying $T$. Recall that $T'$ was described in terms of the sets $T'(n)$ for $n \geq 0$ and the compositions, substitutions, and identities. From this description, the hom sets are $Mor_{T'}(m, n) = \prod_{j=1}^{n} T'(m)$. There is a map of theories $T' \to T$ which gives the theory structure on $T$. Let the underlying 1-category of the 2-category $\mathcal{T}$ be $T'$. For $v, w \in \mathcal{T}(n) = Mor_{\mathcal{T}}(n, 1)$ we define a unique iso 2-cell between $v$ and $w$ if $v$ and $w$ map to the same element of $T(n)$ under the map of theories $T' \to T$. Otherwise there is no 2-cell between $v$ and $w$. With these definitions, the only 2-cell between $w$ and $w$ is the identity and the vertical composition of 2-cells is uniquely defined. Thus $\mathcal{T}(n)$ is a category. Next define $Mor_{\mathcal{T}}(m, n)$ to be the product category $\prod_{j=1}^{n} \mathcal{T}(m)$ for all $m, n \in Obj\, \mathcal{T}$. From this it follows that there is a unique iso 2-cell between $v, w \in Mor_{\mathcal{T}}(m, n)$ if they map to the same element of $Mor_T(m, n)$ and otherwise there is no 2-cell. This uniquely defines the horizontal composition of 2-cells and $\mathcal{T}$ is a 2-category. From the definitions it also follows easily that $n$ is the 2-product of $n$ copies of 1 in $\mathcal{T}$. Hence $\mathcal{T}$ is a theory enriched in groupoids. In [**27**] $\mathcal{T}$ is denoted $(Th(T), G(T))$.

We introduce the notation $c, I, s$ for some of these 2-cells, which breaks the usual convention of labelling 2-cells by lowercase Greek letters. Let

$$c_{w,w_1,\ldots,w_k} : (()_{id_{n_1+\cdots+n_k}}, \gamma(w, w_1, \ldots, w_k)) \Longrightarrow (\gamma, w, w_1, \ldots, w_k)$$

denote the unique 2-cell for $w \in T(k), w_i \in T(n_i), i = 1, \ldots, k$. The $\gamma$ on the right is a generator of the theory of theories while the $\gamma$ on the left is the composition in the theory $T$. The map $id_{n_1+\cdots+n_k}$ is the identity of the object $n_1 + \cdots + n_k$ in the category $\Gamma$ of Definition 6.7. Let

$$I : (()_{id_1}, 1) \Longrightarrow (1, *)$$

where $(()_{id_1}, 1) \in \mathbf{R}_1(1) \times T(1)$ and $(1, *) \in \mathbf{R}_1(1^0) \times T(1)^0$. Here $\mathbf{R}$ denotes the theory of theories in Example 6.29. Let

$$s_{w,f} : (()_{id_n}, w_f) \Longrightarrow (()_f, w)$$

denote the unique 2-cell for $w \in T(m)$ and $f : m \to n$ in $\Gamma$. We call these 2-cells as well as identity 2-cells the *elementary 2-cells*. By the following inductive proof, every other 2-cell in $\mathcal{T}$ can be obtained from these ones and their inverses.

LEMMA 7.11. *Let $\alpha$ be a word in the theory of theories, i.e. $\alpha \in \mathbf{R}_n(\bar{m})$ for some $n \in \mathbb{N}_0$, $\bar{m} = (j_1^{m_1}, \ldots, j_p^{m_p})$, and $m := m_1 + \cdots + m_p$. Then the 2-cell*

$$(()_{id_n}, \alpha(v_1, \ldots, v_m)) \Longrightarrow (\alpha, v_1, \ldots, v_m)$$

*in $\mathcal{T}$ can be expressed as a vertical composition*

$$\sigma_s \odot \sigma_{s-1} \odot \cdots \odot \sigma_1$$

*where each $\sigma_r$ is the result of applying a morphism in $\mathbf{R}$ to a tuple of elementary 2-cells.*

*Proof:* Let $\alpha = \alpha_i \circ \cdots \circ \alpha_1$ where $\alpha_1, \ldots, \alpha_i$ are tuples of generating morphisms in the theory $\mathbf{R}$ of theories such that $i$ is minimal. We induct on $i$. If $i = 1$, then $\alpha$ is a generating morphism for $\mathbf{R}$ and the 2-cell

$$(()_{id_n}, \alpha(v_1, \ldots, v_m)) \Longrightarrow (\alpha, v_1, \ldots, v_m)$$

must be one of $c, I$, or $s$. Now let $i \geq 1$ and suppose the Lemma holds for all words that can be expressed with $i$ terms or less. Suppose $\alpha \in \mathbf{R}_n(\bar{m})$ has an expression with $i + 1$ terms but not does not have an expression with fewer terms. Then $\alpha = \beta \circ (\beta_1, \ldots, \beta_k)$ where $\beta$ is a generating morphism for the theory of theories and $\beta_1, \ldots, \beta_k$ are some words in the theory of theories, each with $i_1, \ldots, i_k \leq i$. Then the 2-cells

$$\begin{aligned}
\varepsilon_1 &: (()_{id}, \beta_1(v_1, \ldots)) \Longrightarrow (\beta_1, v_1, \ldots) \\
\varepsilon_2 &: (()_{id}, \beta_2(\ldots)) \Longrightarrow (\beta_2, \ldots) \\
&\vdots \\
\varepsilon_k &: (()_{id}, \beta_k(\ldots, v_m)) \Longrightarrow (\beta_k, \ldots, v_m)
\end{aligned}$$

can be obtained from elementary 2-cells in the prescribed manner by the induction hypothesis. Here $id$ is generically used to denote any identity morphism in $\Gamma$. Then

$$(()_{id_n}, \alpha(v_1, \ldots, v_m)) = (()_{id_n}, \beta \circ (\beta_1, \ldots, \beta_k)(v_1, \ldots, v_m))$$

$$\Big\Downarrow e$$

$$(\beta, \beta_1(w_1, \ldots), \beta_2(\ldots), \ldots, \beta_k(\ldots, v_m)) = (\beta, (\beta_1, \ldots, \beta_k)(v_1, \ldots, v_m))$$

$$\Big\Downarrow \beta(\varepsilon_1, \ldots, \varepsilon_k)$$

$$(\beta, (\beta_1, w_1, \ldots), (\beta_2, \ldots), \ldots, (\beta_k, \ldots, v_m)) = (\beta \circ (\beta_1, \ldots, \beta_k), v_1, \ldots, v_m)$$

$$\Big\|$$

$$(\alpha, v_1, \ldots, v_m)$$

is also a composition of the prescribed type, where $e$ is an elementary 2-cell. □

LEMMA 7.12. *Let $\alpha$ and $\beta$ be words in the theory of theories. Suppose that there is a 2-cell*

$$(\alpha, v_1, \dots, v_{m_1}) \Longrightarrow (\beta, w_1, \dots, w_{m_2})$$

*in $\mathcal{T}$. Then this 2-cell is a vertical composition of 2-cells obtained from elementary 2-cells and their inverses by applying morphisms in the theory of theories.*

*Proof:* From Lemma 7.11 we have 2-cells

$$\begin{array}{ccc} (\alpha, v_1, \dots, v_{m_1}) & & (\beta, w_1, \dots, w_{m_2}) \\ \Uparrow & & \Uparrow \\ (()_{id}, \alpha(v_1, \dots, v_{m_1})) & = & (()_{id}, \beta(w_1, \dots, w_{m_2})) \end{array}$$

of the prescribed type. We obtain the desired result by inverting the 2-cell on the left. □

THEOREM 7.13. *There is a bijection between the set of small pseudo $T$-algebras and the set of small $\mathcal{T}$-algebras.*

*Proof:* Let $(X, \Phi)$ be a small pseudo $T$-algebra. Define a morphism $\Psi : \mathcal{T} \to End(X)$ of theories enriched in groupoids by the following sequence of functors $\Psi_n : \mathcal{T}(n) \to End(X)(n)$. For notational convenience, the subscript $n$ is usually left off below. For $(\alpha, w_1, \dots, w_\ell) \in \mathcal{T}(n)$ define

$$\Psi(\alpha, w_1, \dots, w_\ell) := \alpha(\Phi(w_1), \dots, \Phi(w_\ell)).$$

For elementary 2-cells, define

$$\Psi(c_{w,w_1,\dots,w_k}) := c_{w,w_1,\dots,w_k}$$
$$\Psi(I) := I$$
$$\Psi(s_{w,f}) := s_{w,f}$$

where the symbols on the right denote the coherence natural isomorphisms from the pseudo $T$-algebra structure.

If $\alpha$ is a word in the theory of theories and $\varepsilon_1, \dots, \varepsilon_k$ are elementary 2-cells, then

$$\Psi(\alpha(\varepsilon_1, \dots, \varepsilon_k)) := \alpha(\Psi(\varepsilon_1), \dots, \Psi(\varepsilon_k)).$$

This is well defined, because if $\alpha(\varepsilon_1, \dots, \varepsilon_k) = \beta(\varepsilon_1, \dots, \varepsilon_k)$ with $\varepsilon_1, \dots, \varepsilon_k$ elementary, then $\alpha = \beta$.

Consider the 2-cell

$$(()_{id_n}, \alpha(v_1, \dots, v_m)) \Longrightarrow (\alpha, v_1, \dots, v_m)$$

for some $\alpha \in \mathbf{R}_n(\overline{m})$. By the above lemma, the word $\alpha$ can be expressed in the form $\sigma_s \odot \cdots \odot \sigma_1$ where each $\sigma_r$ is obtained from a tuple of elementary 2-cells by applying a morphism in $\mathbf{R}$. Define

$$\Psi(\sigma_s \odot \cdots \odot \sigma_1) := \Psi(\sigma_s) \odot \cdots \odot \Psi(\sigma_1)$$

where each $\Psi(\sigma_r)$ is defined as in the previous paragraph. To see that this is well defined, suppose $\sigma_s \odot \cdots \odot \sigma_1 = \sigma'_{s'} \odot \cdots \odot \sigma'_1$ where each $\sigma'_{r'}$ is obtained from a tuple of elementary 2-cells by applying a morphism in $\mathbf{R}$. Such a sequence gives rise to an expression $\alpha = \alpha'_{s'} \circ \cdots \circ \alpha'_1$ where $\alpha'_1, \ldots, \alpha'_{s'}$ are tuples of generating morphisms. Let $\alpha = \alpha_s \circ \cdots \circ \alpha_1$ be the expression that arose from $\sigma_s \odot \cdots \odot \sigma_1$. It suffices to consider the case

$$\alpha = \alpha_4 \circ \alpha_3 \circ \alpha_2 \circ \alpha_1 = \alpha_4 \circ \alpha'_3 \circ \alpha'_2 \circ \alpha_1$$

with $\alpha_3 \circ \alpha_2 = \alpha'_3 \circ \alpha'_2$ because $\alpha'_{s'} \circ \cdots \circ \alpha'_1$ can be obtained from $\alpha_s \circ \cdots \circ \alpha_1$ by a finite number of applications of the relations in the theory of theories. Then we have the following diagram, whose vertical columns are $\Psi(\sigma_4 \odot \sigma_3 \odot \sigma_2 \odot \sigma_1)$ and $\Psi(\sigma_4 \odot \sigma'_3 \odot \sigma'_2 \odot \sigma_1)$ respectively.

$$\begin{array}{ccc}
\Phi(\alpha_4 \circ \alpha_3 \circ \alpha_2 \circ \alpha_1(\bar{w})) & = & \Phi(\alpha_4 \circ \alpha'_3 \circ \alpha'_2 \circ \alpha_1(\bar{w})) \\
{\scriptstyle \varepsilon_4(\alpha_3 \circ \alpha_2 \circ \alpha_1(\bar{w}))} \Downarrow & & \Downarrow {\scriptstyle \varepsilon_4(\alpha'_3 \circ \alpha'_2 \circ \alpha_1(\bar{w}))} \\
\alpha_4 \Phi(\alpha_3 \circ \alpha_2 \circ \alpha_1(\bar{w})) & = & \alpha_4 \Phi(\alpha'_3 \circ \alpha'_2 \circ \alpha_1(\bar{w})) \\
{\scriptstyle \alpha_4(\varepsilon_3(\alpha_2 \circ \alpha_1(\bar{w})))} \Downarrow & & \Downarrow {\scriptstyle \alpha_4(\varepsilon'_3(\alpha'_2 \circ \alpha_1(\bar{w})))} \\
\alpha_4 \circ \alpha_3 \Phi(\alpha_2 \circ \alpha_1(\bar{w})) & & \alpha_4 \circ \alpha'_3 \Phi(\alpha'_2 \circ \alpha_1(\bar{w})) \\
{\scriptstyle \alpha_4 \circ \alpha_3(\varepsilon_2(\alpha_1(\bar{w})))} \Downarrow & & \Downarrow {\scriptstyle \alpha_4 \circ \alpha'_3(\varepsilon'_2(\alpha_1(\bar{w})))} \\
\alpha_4 \circ \alpha_3 \circ \alpha_2 \Phi(\alpha_1(\bar{w})) & = & \alpha_4 \circ \alpha'_3 \circ \alpha'_2 \Phi(\alpha_1(\bar{w})) \\
{\scriptstyle \alpha_4 \circ \alpha_3 \circ \alpha_2(\varepsilon_1(\bar{w}))} \Downarrow & & \Downarrow {\scriptstyle \alpha_4 \circ \alpha'_3 \circ \alpha'_2(\varepsilon_1(\bar{w}))} \\
\alpha_4 \circ \alpha_3 \circ \alpha_2 \circ \alpha_1 \Phi(\bar{w}) & = & \alpha_4 \circ \alpha'_3 \circ \alpha'_2 \circ \alpha_1 \Phi(\bar{w})
\end{array}$$

Here $\varepsilon_i$ denotes the tuple of elementary 2-cells needed to bring $\alpha_i$ past $\Phi$. The inner square commutes because of the coherence diagrams. The top and bottom squares commute because $\alpha_3 \circ \alpha_2 = \alpha'_3 \circ \alpha'_2$. Hence

$$\Psi(\sigma_4 \odot \sigma_3 \odot \sigma_2 \odot \sigma_1) = \Psi(\sigma_4 \odot \sigma'_3 \odot \sigma'_2 \odot \sigma_1)$$

and $\Psi$ is well defined on any 2-cell of the form

$$((\,)_{id_n}, \alpha(v_1, \ldots, v_m)) \Longrightarrow (\alpha, v_1, \ldots, v_m).$$

Next we must define $\Psi$ on 2-cells of the form

$$(\alpha, v_1, \ldots, v_{m_1}) \Longrightarrow (\beta, w_1, \ldots, w_{m_2}).$$

According to Lemma 7.11 we have 2-cells

$$\begin{array}{ccc}
(\alpha, v_1, \ldots, v_{m_1}) & & (\beta, w_1, \ldots, w_{m_2}) \\
{\scriptstyle \mu} \Uparrow & & \Uparrow {\scriptstyle \nu} \\
((\,)_{id}, \alpha(v_1, \ldots, v_{m_1})) & = & ((\,)_{id}, \beta(w_1, \ldots, w_{m_2}))
\end{array}$$

on which $\Psi$ is already defined. Define

$$\Psi(\nu \odot \mu^{-1}) := \Psi(\nu) \odot \Psi(\mu)^{-1}.$$

To see that this is well defined, suppose

$$\sigma_s \odot \cdots \odot \sigma_1 : (\alpha, v_1, \ldots, v_{m_1}) \Longrightarrow (\beta, w_1, \ldots, w_{m_2})$$

is another expression where each $\sigma_r$ is obtained by applying a morphism in $\mathbf{R}$ to a tuple of elementary 2-cells or their inverses. Then

$$\begin{aligned} \Psi(\nu) &= \Psi(\sigma_s \odot \cdots \odot \sigma_1 \odot \mu) \\ \Psi(\nu) &= \Psi(\sigma_s \odot \cdots \odot \sigma_1) \odot \Psi(\mu) \\ \Psi(\nu) \odot \Psi(\mu^{-1}) &= \Psi(\sigma_s \odot \cdots \odot \sigma_1) \\ \Psi(\nu \odot \mu^{-1}) &= \Psi(\sigma_s \odot \cdots \odot \sigma_1) \end{aligned}$$

and $\Psi$ is well defined on 2-cells.

By construction $\Psi_n : \mathcal{T}(n) \to End(X)(n)$ is a functor and it preserves $\gamma, ()_g$, and $(1, *) = 1$. Hence $X$ is a $\mathcal{T}$-algebra with structure maps given by $\Psi$. This procedure $\Phi \mapsto \Psi$ defines a map

$$\text{Pseudo } T\text{-Algebras} \to \mathcal{T}\text{-Algebras}.$$

Now we define a map

$$\mathcal{T}\text{-Algebras} \to \text{Pseudo } T\text{-Algebras}.$$

Let $(X, \Psi)$ be a $\mathcal{T}$-algebra. Then define natural isomorphisms

$$\begin{aligned} c_{w,w_1,\ldots,w_k} &:= \Psi(c_{w,w_1,\ldots,w_k}) \\ I &:= \Psi(I) \\ s_{w,f} &:= \Psi(s_{w,f}) \end{aligned}$$

where the symbols $c, I, s$ on the right are 2-cells in $\mathcal{T}$. Also define

$$\Phi_n(w) := \Psi_n(()_{id_n}, w)$$

for $w \in T(n)$. Then the coherence diagrams are satisfied because $\Psi_n : \mathcal{T}(n) \to End(X)(n)$ is a functor for every $n$ and $\Psi$ preserves $\gamma, ()_g$, and 1.

We can easily check that the two procedures are inverse to one another and that they define a bijection. $\square$

Next we can define a 2-monad $C : Cat \to Cat$ like on page 52. Define a 2-functor $C$ by

$$CX := \frac{(\bigcup_{n \geq 0}(\mathcal{T}(n) \times X^n))}{\Gamma}$$

for any small category $X$. We can similarly define 2-natural transformations $\eta : 1_{Cat} \Rightarrow C$ and $\mu : C^2 \Rightarrow C$.

THEOREM 7.14. *Let $\mathcal{C}_C$ denote the 2-category of small strict $C$-algebras, pseudo morphisms, and 2-cells. Let $\mathcal{C}_T$ denote the 2-category of small pseudo $T$-algebras. Then $\mathcal{C}_C$ and $\mathcal{C}_T$ are 2-equivalent.*

*Proof:* The small $C$-algebras are precisely the small $\mathcal{T}$-algebras by a proof similar to Theorem 6.23. But by the previous theorem, the small $\mathcal{T}$-algebras are precisely the pseudo $T$-algebras. To see that the morphisms of the 2-categories $\mathcal{C}_C$ and $\mathcal{C}_T$ are the same, one must compare the coherence isos of the morphisms. They are related by

$$\rho^C_{(\alpha,w_1,\ldots,w_k)\times(\bar{x})} = \alpha(\rho^T_{w_1}, \ldots, \rho^T_{w_k})(\bar{x}).$$

In diagram (1) of Definition 7.4 the right vertical composition can be replaced by the appropriate component of $\rho^C$ by the composition coherence diagram for coherence isos of pseudo morphisms of $C$-algebras. Then (1) commutes by naturality of $\rho^C$. In (2) of Definition 7.4, the right vertical equality can be replaced by the appropriate component of $\rho^C$ by the unit coherence diagram for coherence isos of pseudo morphisms of $C$-algebras. Then (2) commutes by the naturality of $\rho^C$. Diagram (3) commutes by the naturality of $\rho^C$. The 2-cells of the 2-categories $\mathcal{C}_C$ and $\mathcal{C}_T$ are also the same.

Finally, the 2-equivalence of Theorem 6.46 yields the desired 2-equivalence. □

Power's Theorem 5.3 in [**44**] states that the 2-category of strict $C$-algebras, pseudo morphisms and 2-cells is biequivalent to the 2-category of strict $\mathcal{T}$-algebras, pseudo morphisms, and 2-cells where $\mathcal{T}$ is a theory enriched in categories and $C$ is the corresponding 2-monad in his construction. Power's theorem differs from the above Theorem 7.14 in several regards. Theorem 7.14 above uses strict $C$-algebras to describe pseudo $T$-algebras, where $T$ is a usual theory. Theorem 7.14 also has a 2-equivalence rather than a biequivalence.

Theorem 7.15 states part of Theorem 2.6 from [**9**].

THEOREM 7.15. *(Blackwell, Kelly, Power) Let $C$ be a 2-monad. Then the 2-category of small strict $C$-algebras, pseudo morphisms, and 2-cells of pseudo morphisms admits strictly weighted pseudo limits of strict 2-functors.*

We conclude the following completeness theorem from 7.15.

THEOREM 7.16. *Let $T$ be a theory. Then the 2-category of pseudo $T$-algebras admits strictly weighted pseudo limits of strict 2-functors.*

*Proof:* A 2-equivalence of 2-categories preserves weighted pseudo limits because it admits a left 2-adjoint. Then the result follows from the previous two theorems. □

CHAPTER 8

# Weighted Pseudo Limits in the 2-Category of Pseudo $T$-Algebras

In this chapter we show that the 2-category of pseudo $T$-algebras introduced in Chapter 7 admits weighted pseudo limits. In Chapter 5 we proved that the 2-category of small categories admits weighted pseudo limits in Theorem 5.1, Lemma 5.15, and Theorem 5.16. We modify the proofs in Chapter 5 to obtain Theorem 8.1, Lemma 8.11, and Theorem 8.12. Let $\mathcal{C}$ denote the 2-category of small pseudo $T$-algebras in this chapter. The existence of cotensor products in $\mathcal{C}$ allows us to conclude in Theorem 8.12 that $\mathcal{C}$ admits weighted pseudo limits from a theorem of Street. This result is more general than Theorem 7.16 because it allows the functors to be pseudo. The proof in this chapter for pseudo limits is also constructive, whereas Theorem 7.16 is not.

THEOREM 8.1. *The 2-category $\mathcal{C}$ of small pseudo $T$-algebras admits pseudo limits.*

*Proof:* Let $\mathcal{J}$ be a small 1-category and $F : \mathcal{J} \to \mathcal{C}$ a pseudo functor. Let $\mathbf{1}$ denote the terminal object of the 2-category of small categories as in Theorem 5.1. Let $U$ denote the forgetful 2-functor from the 2-category $\mathcal{C}$ of pseudo $T$-algebras to the 2-category of small categories. The candidate for the pseudo limit of $F$ is $L := PseudoCone(\mathbf{1}, U \circ F)$ as before. Note that these are pseudo cones into the 2-category of small categories, not into the 2-category of pseudo $T$-algebras. We define $\pi : \Delta_L \Rightarrow F$ as in Theorem 5.1. We must show that $L$ has the structure of a pseudo $T$-algebra, that $\pi$ is a pseudo natural transformation to $F$, and that $L$ and $\pi$ are universal. These proofs will draw on the analogous results for the pseudo limit of $U \circ F$.

LEMMA 8.2. *The small category $L$ admits a pseudo $T$-algebra structure.*

*Proof:* We first make the identification of the categories $P$ and $L$ as in Remarks 5.4 and 5.5. Let $\eta^\ell = (a_i^\ell)_i \times (\varepsilon_f^\ell)_f \in Obj\ L$ and $(\xi_i^\ell)_i \in Mor\ L$ for $1 \leq \ell \leq n$ and $w \in T(n)$. We denote the structure maps of the pseudo $T$-algebra $Fi = A_i$ by $\Phi_i$ for all $i \in Obj\ \mathcal{J}$. Let $a_i := \Phi_i(w)(a_i^1, \ldots, a_i^n)$ and $\varepsilon_f := \Phi_{Tf}(w)(\varepsilon_f^1, \ldots, \varepsilon_f^n) \circ \rho_w^{Ff}(a_{Sf}^1, \ldots, a_{Sf}^n) : Ff(a_{Sf}) \to a_{Tf}$ as well as $\xi_i := \Phi_i(w)(\xi_i^1, \ldots, \xi_i^n)$. Then the structure maps of the pseudo $T$-algebra $L$ are defined by $\Phi(w)(\eta^1, \ldots, \eta^n) := (a_i)_i \times (\varepsilon_f)_f$ and $\Phi(w)((\xi_i^1)_i, \ldots, (\xi_i^n)_i) := (\xi_i)_i$. We must verify that these outputs belong to $L$.

We claim that $(a_i)_i \times (\varepsilon_f)_f \in Obj\ L$. We prove this by verifying the coherences in Remarks 5.4 and 5.5 for a fixed word $w \in T(2)$. To avoid cumbersome notation, we write $+$ for $\Psi(w)$ for any structure map $\Psi$. The verification for a general word is the same. We abbreviate $\rho_w^H$ as $\rho^H$ for any morphism $H$ of pseudo $T$-algebras.

The only word appearing in the following diagrams is $w$, so there is no ambiguity. Let $\gamma_{f,g} := \gamma^F_{f,g}$ and $\delta_j := \delta^F_j$. First we show that for all $j \in Obj\, \mathcal{J}$ the diagram.

(8.1)
$$\begin{array}{ccc} a_j & \xrightarrow{\delta_{j*}(a_j)} & F1_j(a_j) \\ & \searrow^{1_{a_j}} & \downarrow{\scriptstyle \varepsilon_{1_j}} \\ & & a_j \end{array}$$

commutes where $a_j = a^1_j + a^2_j$ and $\varepsilon_{1_j} = (\varepsilon^1_{1_j} + \varepsilon^2_{1_j}) \circ \rho^{F1_j}(a^1_j, a^2_j)$ as defined above. After writing this diagram out we get

$$\begin{array}{ccc} a^1_j + a^2_j & \xrightarrow{\delta_{j*}(a^1_j + a^2_j)} & F1_j(a^1_j + a^2_j) \\ {\scriptstyle \rho^{1_{Fj}}(a^1_j, a^2_j) = 1_{a^1_j + a^2_j}}\downarrow & & \downarrow{\scriptstyle \rho^{F1_j}(a^1_j, a^2_j)} \\ a^1_j + a^2_j & \xrightarrow{\delta_{j*}(a^1_j) + \delta_{j*}(a^2_j)} & F1_j(a^1_j) + F1_j(a^2_j) \\ {\scriptstyle 1_{a^1_j + a^2_j}}\downarrow & & \downarrow{\scriptstyle \varepsilon^1_{1_j} + \varepsilon^2_{1_j}} \\ a^1_j + a^2_j & \xrightarrow[1_{a^1_j + a^2_j}]{} & a^1_j + a^2_j \end{array}$$

where the top horizontal arrow is $\delta_{j*}(a_j)$ and the right vertical composition is $\varepsilon_{1_j}$ by definition. The top square commutes because $\delta_{j*} : 1_{Fj} \Rightarrow F1_j$ is a 2-cell in the 2-category $\mathcal{C}$. The bottom square commutes because $+$ is a functor and $\varepsilon^\ell_{1_j} \circ \delta_{j*}(a^\ell_j) = 1_{a^\ell_j}$ for $\ell = 1, 2$. Hence (8.1) commutes. Next we show that for all $i \xrightarrow{f} j \xrightarrow{g} k$ in $\mathcal{J}$ the diagram

(8.2)
$$\begin{array}{ccc} Fg \circ Ff(a_i) & \xrightarrow{\gamma_{f,g}(a_i)} & F(g \circ f)(a_i) \\ {\scriptstyle Fg(\varepsilon_f)}\downarrow & & \downarrow{\scriptstyle \varepsilon_{g \circ f}} \\ Fg(a_j) & \xrightarrow[\varepsilon_g]{} & a_k \end{array}$$

commutes where $\varepsilon_f = \varepsilon^1_f + \varepsilon^2_f$ *etc.* After writing out this diagram we get the diagram below whose outermost square is (8.2). The upper left triangle commutes by the definition of composition for morphisms of pseudo $T$-algebras. The upper right quadrilateral commutes because $\gamma_{f,g} : Fg \circ Ff \Rightarrow F(g \circ f)$ is a 2-cell in the 2-category of pseudo $T$-algebras. The lower left square commutes because $\rho^{Fg} : Fg(+) \Rightarrow Fg + Fg$ is a natural transformation. The bottom right square commutes because $+$ is a functor and $\varepsilon^\ell_g \circ (Fg(\varepsilon^\ell_f)) = \varepsilon^\ell_{g \circ f} \circ \gamma_{f,g}(a^\ell_i)$ for $\ell = 1, 2$. Thus all four inner diagrams commute and (8.2) commutes. Thus both coherences in Remark 5.4 are satisfied and $\eta^1 + \eta^2 = (a_i)_i \times (\varepsilon_f)_f$ is an object of $L$.

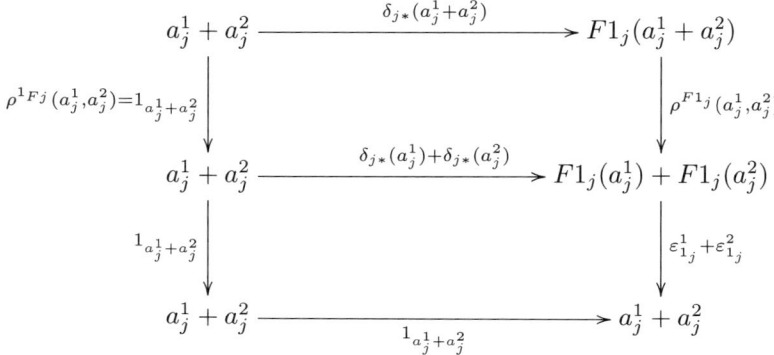

$$\xymatrix{
Fg\circ Ff(a_i^1+a_i^2) \ar[rr]^{\gamma_{f,g}(a_i^1+a_i^2)} \ar[d]_{Fg(\rho^{Ff}(a_i^1,a_i^2))} \ar[dr]^{\rho^{Fg\circ Ff}(a_i^1,a_i^2)} & & F(g\circ f)(a_i^1+a_i^2) \ar[d]^{\rho^{F(g\circ f)}(a_i^1,a_i^2)} \\
Fg(Ff(a_i^1)+Ff(a_i^2)) \ar[r]_{\rho^{Fg}(Ff(a_i^1),Ff(a_i^2))} \ar[d]_{Fg(\varepsilon_j^1+\varepsilon_j^2)} & Fg\circ Ff(a_i^1)+Fg\circ Ff(a_i^2) \ar[r]_{\gamma_{f,g}(a_i^1)+\gamma_{f,g}(a_i^2)} \ar[d]^{Fg(\varepsilon_f^1)+Fg(\varepsilon_f^2)} & F(g\circ f)(a_i^1)+F(g\circ f)(a_i^2) \ar[d]^{\varepsilon_{g\circ f}^1+\varepsilon_{g\circ f}^2} \\
Fg(a_j^1+a_j^2) \ar[r]_{\rho^{Fg}(a_j^1,a_j^2)} & Fg(a_j^1)+Fg(a_j^2) \ar[r]_{\varepsilon_g^1+\varepsilon_g^2} & a_k^1+a_k^2
}$$

We claim that $(\xi_i^1)_i + (\xi_i^2) = (\xi_i)_i$ is a morphism in $L$ where $(\xi_i^1)_i : (a_i^1)_i \times (\varepsilon_f^1)_f \to (b_i^1)_i \times (\zeta_f^1)_f$ and $(\xi_i^2)_i : (a_i^2)_i \times (\varepsilon_f^2)_f \to (b_i^2)_i \times (\zeta_f^2)_f$ are morphisms in $L$. In other words we must show that

$$\xymatrix{
Ff(a_i) \ar[r]^{\varepsilon_f} \ar[d]_{Ff(\xi_i)} & a_j \ar[d]^{\xi_j} \\
Ff(b_i) \ar[r]_{\zeta_f} & b_j
} \tag{8.3}$$

commutes for all morphisms $f : i \to j$ in $\mathcal{J}$, where $a_i = a_i^1 + a_i^2$ *etc.* If we write out the diagram we get

$$\xymatrix{
Ff(a_i^1+a_i^2) \ar[r]^{\rho^{Ff}(a_i^1,a_i^2)} \ar[d]_{Ff(\xi_i^1+\xi_i^2)} & Ff(a_i^1)+Ff(a_i^2) \ar[r]^{\varepsilon_f^1+\varepsilon_f^2} \ar[d]_{Ff(\xi_i^1)+Ff(\xi_i^2)} & a_j^1+a_j^2 \ar[d]^{\xi_j^1+\xi_j^2} \\
Ff(b_i^1+b_i^2) \ar[r]_{\rho^{Ff}(b_i^1,b_i^2)} & Ff(b_i^1)+Ff(b_i^2) \ar[r]_{\zeta_f^1+\zeta_f^2} & b_j^1+b_j^2
}$$

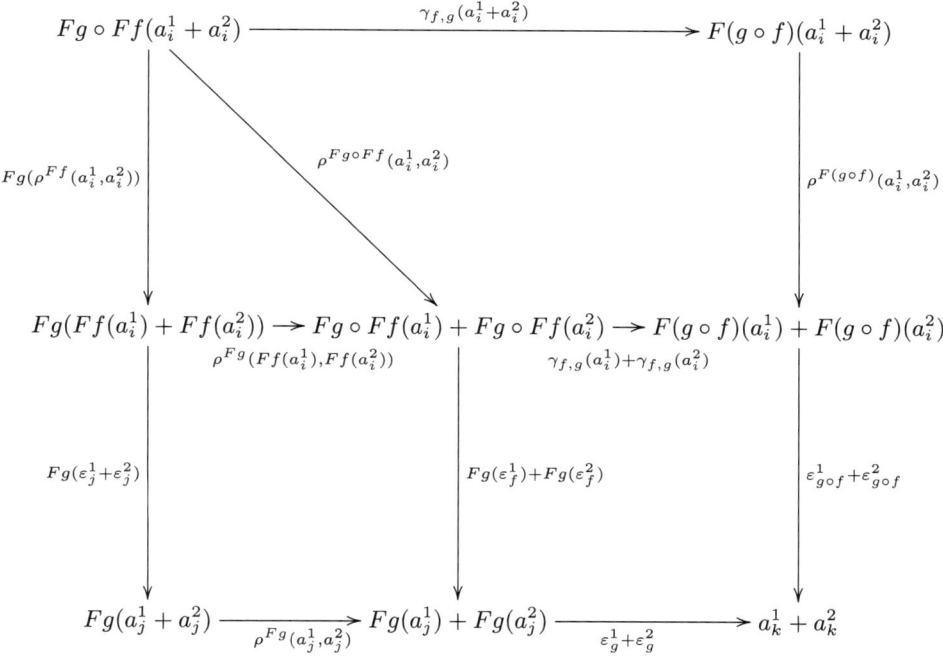

where the outermost square is (8.3). The square on the left commutes because $\rho^{Ff} : Ff(+) \Rightarrow Ff + Ff$ is a natural transformation. The right square commutes because the diagram

$$\begin{array}{ccc} Ff(a_i^\ell) & \xrightarrow{\varepsilon_f^\ell} & a_j \\ {\scriptstyle Ff(\xi_i^\ell)}\downarrow & & \downarrow{\scriptstyle \xi_j^\ell} \\ Ff(b_i) & \xrightarrow[\zeta_f^\ell]{} & b_j^\ell \end{array}$$

commutes for $\ell = 1, 2$ and because $+$ is a functor. Hence $(\xi_i^1)_i + (\xi_i^2) = (\xi_i)_i$ is a morphism in $L$. Thus $\Phi(w) : L \times L \to L$.

The map $\Phi(w)$ preserves compositions and identities because the individual components do. Thus $\Phi(w) : L \times L \to L$ is a functor. The same argument works for words in $T(n)$ for all $n \in \mathbb{N}$. Thus $\Phi$ defines structure maps to make the small category $L$ into a pseudo $T$-algebra.

We define the coherence isos for $\Phi$ to be those maps which have the coherence isos of $\Phi_i$ in the $i$-th component. We can prove that they are morphisms of the category $L$, *i.e.* satisfy the diagram in Remark 5.5, by using the coherence diagrams of $\rho$ with the respective coherence iso as well as the naturality of the individual components. The coherence isos for $\Phi$ are natural because they are natural in each component. The coherence isos for $\Phi$ satisfy the coherence diagrams because the individual components do. Thus $L$ is a pseudo $T$-algebra with structure maps $\Phi$.

□

LEMMA 8.3. *The map $\pi : \Delta_L \Rightarrow F$ is a pseudo natural transformation with coherence iso 2-cells given by $\tau$.*

*Proof:* It is clear from the work on the small category case in Chapter 5 that $\pi$ is a pseudo natural transformation when we forget all the pseudo $T$-algebra structures. Therefore it suffices to show that $\pi_j : L \to Fj$ is a morphism of pseudo $T$-algebras for all $j \in Obj\ \mathcal{J}$ and that $\tau_{i,j}(f) : Ff \circ \pi_i \Rightarrow \pi_j$ is a 2-cell in the 2-category of pseudo $T$-algebras for all morphisms $f : i \to j$ in $\mathcal{J}$.

Let $j \in Obj\ \mathcal{J}$. Then $\pi_j : L \to Fj$ is a functor. We abbreviate $\Phi(w)$ for $w \in T(2)$ by $+$ as above. Then for $\eta^\ell = (a_i^\ell)_i \times (\varepsilon_f^\ell)_f \in Obj\ L$ for $\ell = 1, 2$ we have

$$\begin{aligned} \pi_j(\eta^1 + \eta^2) &= \pi_j((a_i^1 + a_i^2)_i \times ((\varepsilon_f^1 + \varepsilon_f^2) \circ \rho_w^{Ff}(a_{Sf}^1, a_{Sf}^2))_f) \\ &= a_j^1 + a_j^2 \\ &= \pi_j(\eta^1) + \pi_j(\eta^2). \end{aligned}$$

The same calculation works for words in $T(n)$ for all $n \in \mathbb{N}$. We conclude that $\pi_j$ commutes with the structure maps for the pseudo $T$-algebra structure. If we take $\rho_w^{\pi_j} = i_{\pi_j} \circ i_{\Phi(w)}$ then $\pi_j$ is a morphism of pseudo $T$-algebras for all $j \in \mathcal{J}$.

Let $f : i \to j$ be a morphism in $L$. To show that $\tau_{i,j}(f)$ is a 2-cell, we must show that the diagram

(8.4)

$$\begin{array}{ccc} Ff \circ \pi_i \circ \Phi(w) & \xRightarrow{\tau_{i,j}(f) * i_{\Phi(w)}} & \pi_j \circ \Phi(w) \\ \Big\Downarrow \rho_w^{Ff \circ \pi_i} & & \Big\Downarrow \rho_w^{\pi_j} \\ \Phi_j(w) \circ (Ff \circ \pi_i, \ldots, Ff \circ \pi_i) & \xRightarrow[i_{\Phi_j(w)} * (\tau_{i,j}(f), \ldots, \tau_{i,j}(f))]{} & \Phi_j(w) \circ (\pi_j, \ldots, \pi_j) \end{array}$$

commutes for all words $w$. Recalling that $\tau_{i,j}(f)_\eta := \tau^\eta_{i,j}(f)$ and evaluating the diagram on $(\eta^1, \eta^2)$ where $\eta^\ell = (a^\ell_i)_i \times (\varepsilon^\ell_f)_f \in Obj\ L$ for $\ell = 1, 2$ gives

$$\begin{array}{ccc} Ff(a^1_i + a^2_i) & \xrightarrow{(\varepsilon^1_f + \varepsilon^2_f) \circ \rho^{Ff}(a^1_i, a^2_i)} & a^1_j + a^2_j \\ \Big\downarrow \rho^{Ff}(a^1_i, a^2_i) & & \Big\downarrow 1_{a^1_j + a^2_j} \\ Ff(a^1_i) + Ff(a^2_i) & \xrightarrow[\varepsilon^1_f + \varepsilon^2_f]{} & a^1_j + a^2_j \end{array}$$

which obviously commutes. Hence $\tau_{i,j}(f)$ is a 2-cell in the 2-category of pseudo $T$-algebras for all $f : i \to j$ and $\pi$ is a pseudo natural transformation.

□

Now we must show that the pseudo $T$-algebra $L$ and the pseudo natural transformation $\pi : \Delta_L \Rightarrow F$ are universal in the sense that the functor $\phi : Mor_{\mathcal{C}}(V, L) \to PseudoCone(V, F)$ as defined in the small category case of Chapter 5 is an isomorphism of categories for all objects $V$ of $\mathcal{C}$. In the following, $V$ is a fixed object of the 2-category $\mathcal{C}$ of pseudo $T$-algebras.

LEMMA 8.4. *The map* $\phi : Mor_{\mathcal{C}}(V, L) \to PseudoCone(V, F)$ *is a functor.*

*Proof:* The proof is analogous to the proof for the $\phi$ of the pseudo colimit of small categories in Lemma 4.4. The only difference is that here we have to verify that $\tau_{i,j}(f) * i_b$ is a 2-cell of the 2-category $\mathcal{C}$ of pseudo $T$-algebras for any morphism $b : V \to L$ as in the comments just before Lemma 4.4. But that is immediate because $i_b$ is obviously a 2-cell and the horizontal composition of 2-cells is again a 2-cell. □

Now we construct a functor $\psi : PseudoCone(V, F) \to Mor_{\mathcal{C}}(V, L)$ that is inverse to $\phi$. First we define $\psi$ for objects, then for morphisms. Finally we verify that it is a functor and inverse to $\phi$. The next two lemmas define a morphism $\psi(\pi') : V \to L$ in $\mathcal{C}$ for any object $\pi'$ of $PseudoCone(V, F)$.

LEMMA 8.5. *Let* $\pi' : \Delta_V \Rightarrow F$ *be a pseudo natural transformation with coherence 2-cells* $\tau'$. *For any fixed* $x \in Obj\ V$ *we have* $\psi(\pi')(x) := b(x) := (\pi'_i(x))_i \times (\tau'_{Sf,Tf}(f)_x)_f$ *is an element of* $Obj\ L$.

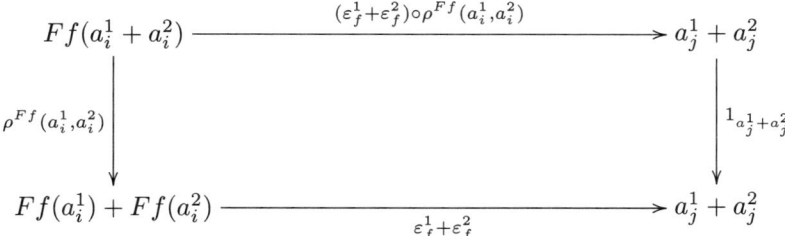

*Proof:* This follows from Lemma 5.7 by forgetting the pseudo $T$-algebra structures. Thus $\psi(\pi')(x) \in Obj\ L$. $\square$

LEMMA 8.6. *Let $\pi' : \Delta_V \Rightarrow F$ be a pseudo natural transformation with coherence 2-cells $\tau'$. Then for any fixed $h \in Mor_V(x,y)$ we have a modification $\psi(\pi')(h) := b(h) := (\pi_i'(h))_i : b(x) \rightsquigarrow b(y)$. This notation means $b(h)_i(*) := \pi_i'(h)$.*

*Proof:* This is exactly the same as the proof of Lemma 5.8 because the pseudo $T$-algebra structure on $L$ makes no additional requirements on the morphisms of the small category $L$. $\square$

LEMMA 8.7. *For any pseudo natural transformation $\pi' : \Delta_V \Rightarrow F$ the map $\psi(\pi') = b : V \to L$ as defined above is a morphism of pseudo $T$-algebras.*

*Proof:* By Lemma 5.9 the map $b : V \to L$ is a functor between the underlying small categories. We define a natural transformation $\rho^b_w$ for $w \in T(2)$. We abbreviate the application of any structure map to $w$ by $+$. Define $\rho^b_w(x_1,x_2) := \rho^b(x_1,x_2) := (\rho^{\pi_i'}(x_1,x_2))_i : b(x_1+x_2) \to b(x_1)+b(x_2)$ for all $x_1,x_2 \in Obj\ V$. We claim that $\rho^b(x_1,x_2)$ is a morphism in $L$. Let $\tau'_{i,j}(f)$ denote the coherence 2-cell of $\pi' : \Delta_V \Rightarrow F$ for $f : i \to j$ in $\mathcal{J}$. Since $\tau'_{i,j}(f) : Ff \circ \pi' \Rightarrow \pi_j'$ is a 2-cell, we know that

$$\begin{array}{ccc}
Ff \circ \pi_i'(x_1+x_2) & \xrightarrow{\ \tau'_{i,j}(f)_{x_1+x_2}\ } & \pi_j'(x_1+x_2) \\
{\scriptstyle \rho^{Ff\circ\pi_i'}(x_1,x_2)}\Big\downarrow & & \Big\downarrow{\scriptstyle \rho^{\pi_j'}(x_1,x_2)} \\
Ff \circ \pi_i'(x_1) + Ff \circ \pi_i'(x_2) & \xrightarrow[\ \tau'_{i,j}(f)_{x_1}+\tau'_{i,j}(f)_{x_2}\ ]{} & \pi_j'(x_1)+\pi_j'(x_2)
\end{array}$$

commutes. Rewriting the left vertical arrow and the bottom arrow gives

$$\begin{array}{ccc}
Ff(\pi_i'(x_1+x_2)) & \xrightarrow{\ \tau'_{i,j}(f)_{x_1+x_2}\ } & \pi_j'(x_1+x_2) \\
{\scriptstyle Ff\rho^{\pi_i'}(x_1,x_2)}\Big\downarrow & & \Big\downarrow{\scriptstyle \rho^{\pi_j'}(x_1,x_2)} \\
Ff(\pi_i'(x_1)+\pi_i'(x_2)) & \xrightarrow[\ (\tau'_{i,j}(f)_{x_1}+\tau'_{i,j}(f)_{x_2})\circ\rho^{Ff}(\pi_i'(x_1),\pi_i'(x_2))\ ]{} & \pi_j'(x_1)+\pi_j'(x_2)
\end{array}$$

which states precisely that $\rho^b(x_1,x_2) = (\rho^{\pi_i'}(x_1,x_2))_i : b(x_1+x_2) \to b(x_1)+b(x_2)$ is a morphism in $L$ by Remark 5.5. The map $\rho^b$ is natural because each component is natural. Hence $\rho^b$ is a natural transformation. If we define $\rho^b_w$ analogously for arbitrary words $w$ of the theory $T$, then the coherences of Definition 7.1 are satisfied because they are satisfied componentwise. Hence $\psi(\pi') = b : V \to L$ is a morphism of pseudo $T$-algebras. $\square$

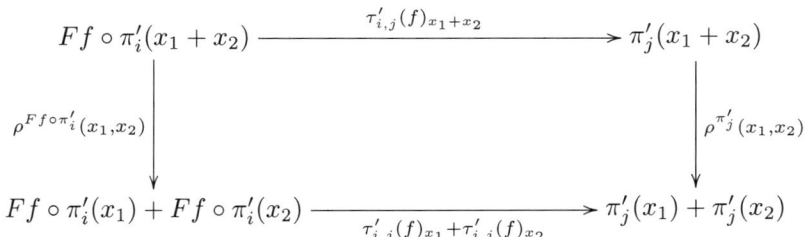

LEMMA 8.8. *Let $\Xi : \alpha \rightsquigarrow \beta$ be a morphism in the category $PseudoCone(V,F)$. Then $\psi(\Xi) : \psi(\alpha) \Rightarrow \psi(\beta)$ defined by $V \ni x \mapsto (\Xi_i(x))_i \in Mor_L(\psi(\alpha)x, \psi(\beta)x)$ is*

*a 2-cell in the 2-category of pseudo $T$-algebras. As in Lemma 5.10, this definition means $\psi(\Xi)(x)_i(*) := \Xi_i(x)$.*

*Proof:* The map $\psi(\Xi)$ is a natural transformation by Lemma 5.10. For all $i \in Obj\ \mathcal{J}$ we have morphisms $\alpha_i, \beta_i : V \to Fi$ and 2-cells $\Xi_i : \alpha_i \Rightarrow \beta_i$. Hence

$$\begin{array}{ccc}
\alpha_i(x_1+x_2) & \xrightarrow{\Xi_i(x_1+x_2)} & \beta_i(x_1+x_2) \\
\Big\downarrow {\scriptstyle \rho^{\alpha_i}(x_1,x_2)} & & \Big\downarrow {\scriptstyle \rho^{\beta_i}(x_1,x_2)} \\
\alpha_i(x_1)+\alpha_i(x_2) & \xrightarrow[\Xi_i(x_1)+\Xi_i(x_2)]{} & \beta_i(x_1)+\beta_i(x_2)
\end{array}$$

commutes. Since these are the components for $\psi(\alpha)(x), \psi(\beta)(x)$, and $\psi(\Xi)(x)$, we see that

$$\begin{array}{ccc}
\psi(\alpha)(x_1+x_2) & \xrightarrow{\psi(\Xi)(x_1+x_2)} & \psi(\beta)(x_1+x_2) \\
\Big\downarrow {\scriptstyle \rho^{\psi(\alpha)}(x_1,x_2)} & & \Big\downarrow {\scriptstyle \rho^{\psi(\beta)}(x_1,x_2)} \\
\psi(\alpha)(x_1)+\psi(\alpha)(x_2) & \xrightarrow[\psi(\Xi)(x_1)+\psi(\Xi)(x_2)]{} & \psi(\beta)(x_1)+\psi(\beta)(x_2)
\end{array}$$

commutes. Similar diagrams hold for arbitrary words $w$ in the theory $T$. Thus $\psi(\Xi)$ is a 2-cell. □

THEOREM 8.9. *The map $\psi : PseudoCone(V, F) \to Mor_{\mathcal{C}}(V, L)$ as defined in the previous lemmas is an inverse functor to $\phi$.*

*Proof:* This follows from the calculations of Theorem 5.11 and Lemmas 5.12 and 5.13. □

LEMMA 8.10. *The pseudo $T$-algebra $L$ with the pseudo cone $\pi : \Delta_L \Rightarrow F$ is a pseudo limit of the pseudo functor $F : \mathcal{J} \to \mathcal{C}$.*

*Proof:* The functor $\phi : Mor_{\mathcal{C}}(V, L) \to PseudoCone(V, F)$ is an isomorphism of categories by the previous lemmas. Since $V$ was an arbitrary object of $\mathcal{C}$ we conclude that $L$ and $\pi$ are universal. □

Thus every pseudo functor $F : \mathcal{J} \to \mathcal{C}$ from a small 1-category $\mathcal{J}$ to the 2-category $\mathcal{C}$ of pseudo $T$-algebras admits a pseudo limit. Hence $\mathcal{C}$ admits pseudo limits. This completes the proof of Theorem 8.1.

□

LEMMA 8.11. *The 2-category $\mathcal{C}$ of small pseudo $T$-algebras admits cotensor products.*

*Proof:* Let $J \in Obj\ Cat$ and let $F$ be a pseudo $T$-algebra. Let $U : \mathcal{C} \to Cat$ be the forgetful functor. Define $P := (UF)^J$, which is the 1-category of 1-functors $J \to UF$. We claim that $P$ has the structure of a pseudo $T$-algebra. Let $\Phi_n$ :

$T(n) \to Functors(F^n, F)$ denote the structure maps for $F$. Define $\Phi_n^P : T(n) \to Functors(P^n, P)$ by

$$\Phi_n^P(w)(p_1, \ldots, p_n)(j) := \Phi_n(w)(p_1(j), \ldots, p_n(j))$$

for $j \in Obj\ J$ and $p_1, \ldots, p_n \in Obj\ P$. Coherence isos are defined analogously. For example, define $s^P_{w,f} : \Phi^P(w_f) \Rightarrow \Phi^P(w)_f$ for $f : m \to n$ on $p_1, \ldots, p_n \in Obj\ P$ as the 1-natural transformation

$$s^P_{w,f}(p_1, \ldots, p_n) : \Phi_n^P(w_f)(p_1, \ldots, p_n) \Longrightarrow \Phi_n^P(w)_f(p_1, \ldots, p_n)$$

which is $s^P_{w,f}(p_1, \ldots, p_n)(j) := s_{w,f}(p_1(j), \ldots, p_n(j))$ for $j \in Obj\ J$. Then all coherence diagrams are satisfied because they are satisfied pointwise. Hence, $P$ has the structure of a pseudo $T$-algebra.

We claim that $P$ is a cotensor product of $J$ and $F$. We use Remark 3.21. Define a functor $\pi : J \to \mathcal{C}(P, F)$ by

$$\pi(j)(p) := p(j)$$
$$\pi(j)(\eta) := \eta(j)$$
$$\pi(g)(p) := p(g)$$

for $j$ an object of $J$, $p$ a functor from $J$ to $UF$, $\eta$ a natural transformation, and $g$ a morphism in $J$. Let $\sigma : J \to \mathcal{C}(C, F)$ be a functor. Define a morphism $b : C \to P$ of pseudo $T$-algebras by

$$b(c)(j) := \sigma(j)(c)$$
$$b(c)(f) := \sigma(f)(c)$$
$$b(m)(j) := \sigma(j)(m)$$

for $c \in Obj\ C$, $j \in Obj\ J$, $f \in Mor\ J$, and $m \in Mor\ C$. Then $b$ is strict and it is the unique morphism $C \to P$ such that $\mathcal{C}(b, F) \circ \pi = \sigma$. A similar argument can be made for 2-cells. Thus $P$ is a cotensor product of $J$ and $F$ with unit $\pi$. □

THEOREM 8.12. *The 2-category $\mathcal{C}$ of small pseudo $T$-algebras admits weighted pseudo limits.*

*Proof:* By Theorem 8.1 it admits pseudo limits, and hence it admits pseudo equalizers. The 2-category $\mathcal{C}$ obviously admits 2-products. By Lemma 8.11 it admits cotensor products. Hence by Theorem 3.22 it admits weighted pseudo limits. □

THEOREM 8.13. *The 2-category $\mathcal{C}$ of small pseudo $T$-algebras admits weighted bilimits.*

*Proof:* It admits weighted pseudo limits and therefore admits weighted bilimits. □

CHAPTER 9

# Biuniversal Arrows and Biadjoints

After studying bilimits and bicolimits, we turn our attention to another type of weakened structure called *biadjoints*. The concept of an adjunction from 1-category theory consists of two functors and a natural bijection between appropriate hom sets. Mac Lane lists several equivalent ways of describing an adjunction in [**39**] on pages 79-86. One of these ways involves a universal arrow for each object of the source category. To weaken these concepts, we replace the functors by pseudo functors, the natural bijection of hom sets by a pseudo natural equivalence of categories, and the universal arrow by a biuniversal arrow. The main goal in this chapter is to prove that a biadjunction can be described via pseudo natural equivalences or via biuniversal arrows. This is the meaning of Theorem 9.16 and Theorem 9.17.

A close result in the literature can be found in Gray's work [**19**]. His concept of *transcendental quasiadjunction* between two 2-functors on page 177 is similar to the concept of biadjunction between two pseudo functors except that the functors in a biadjoint are allowed to be pseudo. Gray remarks on pages 180-181 that a transcendental quasiadjunction gives rise to a certain universal mapping property. The analogous concept for biadjoints is a biuniversal arrow and the appropriate theorem is Theorem 9.16. On page 184 Gray remarks that under certain hypothesis, the universal mapping property gives rise to a quasiadjunction. The biadjoint version of this is Theorem 9.17 in which the starting functor $G$ is allowed to be a pseudo functor.

Kelly phrases a similar result in [**29**] on page 316 in terms of homomorphisms of bicategories and birepresentations. His notion of biadjoint is the same as in this paper, except that we are considering only pseudo functors between 2-categories rather than homomorphisms between bicategories. Kelly's statement is equivalent to 9.17 after an application of Yoneda's Lemma for bicategories. Yoneda's Lemma for bicategories can be found in [**50**].

Street makes an observation on page 121 in [**50**] similar to Theorem 9.17: if each object admits a left bilifting then a left biadjoint exists. The unit for a left bilifting is the biuniversal arrow of Theorem 9.17.

MacDonald and Stone also have a weakened notion of adjunction in [**41**] called *soft adjunction*. In that article they consider strict 2-functors and natural adjunctions between hom categories. They prove theorems about the universality concepts that arise in such a context.

We follow Mac Lane's presentation of adjoints except we account for the 2-cells. The notation in this study is analogous to the notation in Mac Lane's book. Recall the definition of a universal arrow and its uniqueness.

DEFINITION 9.1. Let $S : D \to C$ be a functor between 1-categories and $c \in Obj\ C$. Then an object $r \in Obj\ D$ and a morphism $u \in Mor_C(c, Sr)$ are a *universal arrow from $c$ to $S$* if for every $d \in Obj\ D$ and every $f \in Mor_C(c, Sd)$ there exists a unique morphism $f' \in Mor_D(r,d)$ such that $Sf' \circ u = f$. Pictorially this means for every $d$ and every $f$ as above, there exists a unique $f'$ making

$$\begin{array}{ccc} c \xrightarrow{u} Sr & & r \\ \parallel \quad \downarrow Sf' & & \downarrow f' \\ c \xrightarrow{f} Sd & & d \end{array}$$

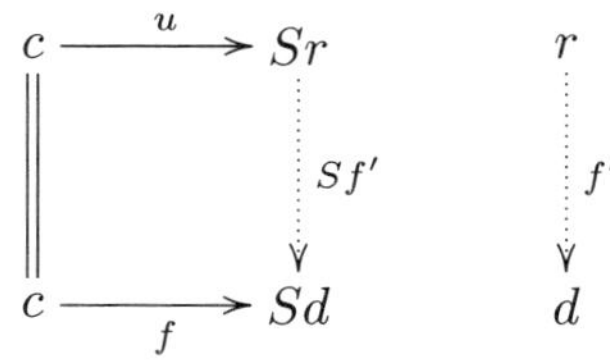

commute. This is equivalent to saying the assignment $f' \mapsto Sf' \circ u$, $Mor_D(r,d) \to Mor_C(c, Sd)$ is a bijection of hom sets for every fixed $d \in Obj\ D$.

LEMMA 9.2. *Let $u : c \to Sr$ and $u' : c \to Sr'$ be universal arrows from the object $c$ to the functor $S$. Then there exists a unique morphism $f' : r \to r'$ such that $Sf' \circ u = u'$. Moreover, the morphism $f' : r \to r'$ is an isomorphism.*

*Proof:* There exist unique morphisms $f'$ and $g'$ such that the following diagram commutes.

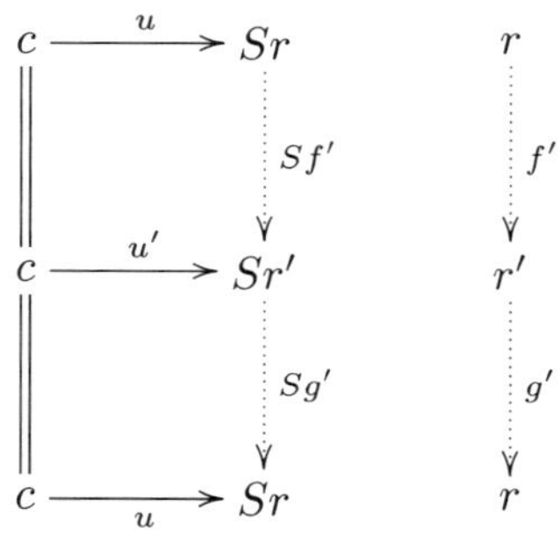

The middle vertical column could be replaced by $S1_r$ to make the outermost rectangle commutative. Hence by the uniqueness we have $g' \circ f' = 1_r$. Similarly we can show that $f' \circ g' = 1_{r'}$. Hence $f'$ is an isomorphism and $Sf' \circ u = u'$. □

Before weakening the concept of universal arrow, we prove a simple lemma that will make it easier to visualize a biuniversal arrow.

LEMMA 9.3. *Let $X \underset{\psi}{\overset{\phi}{\rightleftarrows}} A$ be adjoint functors with unit $\theta : 1_X \Rightarrow \psi \circ \phi$ and counit $\mu : \phi \circ \psi \Rightarrow 1_A$. Suppose that both the unit and the counit are natural isomorphisms. Let $\nu : \phi(x) \to a$ be a morphism in $A$ and $x \in Obj\ X, a \in Obj\ A$. Then there exists a unique morphism $\nu' : x \to \psi(a)$ such that*

$$\begin{array}{ccc} x & \phi(x) \xrightarrow{\nu} a \\ \downarrow \nu' & \downarrow \phi(\nu') \quad \parallel \\ \psi(a) & \phi(\psi(a)) \xrightarrow{\mu(a)} a \end{array}$$

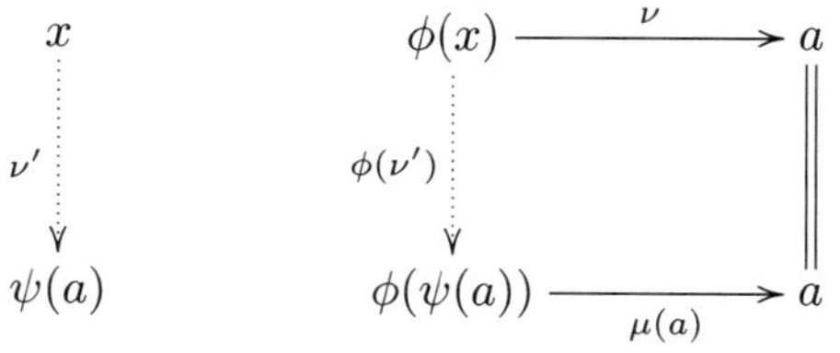

*commutes. Moreover, $\nu'$ is iso if and only if $\nu$ is iso.*

*Proof:* The existence and uniqueness claims follow because $\mu(a)$ is a universal arrow from $\phi$ to $a$. If $\nu'$ is iso, then $\phi(\nu')$ is iso and so is $\nu = \mu(a) \circ \phi(\nu')$ because

$\mu(a)$ is iso by hypothesis. It only remains to show that $\nu'$ is iso if $\nu$ is iso. Suppose $\nu$ is iso. Then $\phi(\nu')$ is iso from the commutivity of the diagram because $\mu(a)$ and $\nu$ are iso. By the naturality of $\theta$ we have

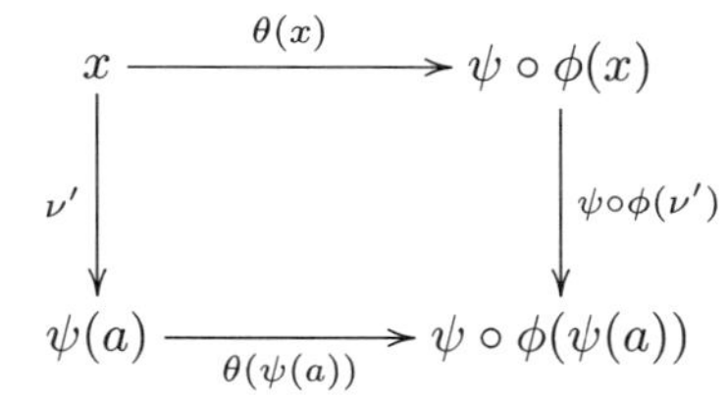


commutes. Then $\nu'$ is iso because $\theta(x), \theta(\psi(a))$, and $\psi(\phi(\nu'))$ are iso. □

To weaken the concept of universal arrow in the context of 2-categories, we replace the bijection of sets above by an equivalence of the appropriate morphism categories.

DEFINITION 9.4. Let $S : \mathcal{D} \to \mathcal{C}$ be a pseudo functor between 2-categories and $C \in Obj\ \mathcal{C}$. Then an object $R \in Obj\ \mathcal{D}$ and a morphism $u \in Mor_{\mathcal{C}}(C, SR)$ are a *biuniversal arrow* from $C$ to $S$ if for every $D \in Obj\ \mathcal{D}$ the functor $\phi : Mor_{\mathcal{D}}(R, D) \to Mor_{\mathcal{C}}(C, SD)$ defined by $f' \mapsto Sf' \circ u$ and $\gamma \mapsto S\gamma * i_u$ is an equivalence of categories.

We suppressed the dependence of $\phi$ on $D$ in the notation of the definition. This definition implies that $\phi$ admits a right adjoint $\psi$ such that the counit $\mu : \phi \circ \psi \Rightarrow 1_{Mor_{\mathcal{C}}(C,SD)}$ and unit are natural isomorphisms. Pictorially the definition implies that for every object $D \in Obj\ \mathcal{D}$ and every morphism $f : C \to SD$ in $\mathcal{C}$ there exists an $f'$ and a natural universal 2-cell $\mu(f)$ which is iso (an arrow of the counit) as in the following diagram.

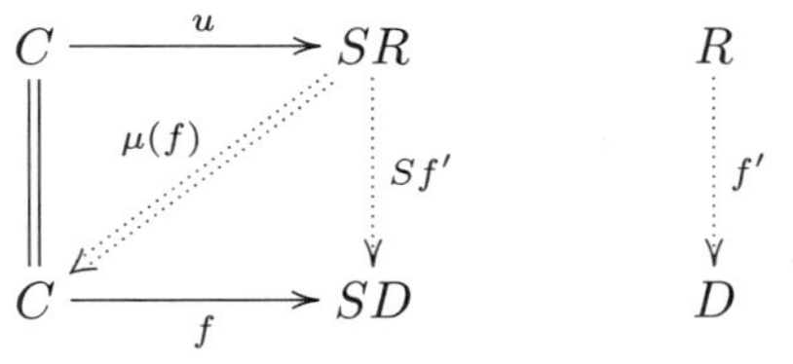


The assignment $\psi : f \mapsto f'$ is functorial and $\mu : \phi \circ \psi \Rightarrow 1_{Mor_{\mathcal{C}}(C,SD)}$ is a natural transformation. This diagram is not equivalent to the definition because it does not express the naturality of the 2-cells, nor does it include the natural isomorphism (the unit) from the identity functor on $Mor_{\mathcal{D}}(R, D)$ to $\psi \circ \phi$. The universality of the 2-cell $\mu(f)$ from the functor $\phi$ to the object $f$ means pictorially that the arrow $f'$ is unique up to 2-cell in the following way. If $\bar{f}' : R \to D$ is an arrow in $\mathcal{D}$ and $\nu$ is a (not necessarily iso) 2-cell as in

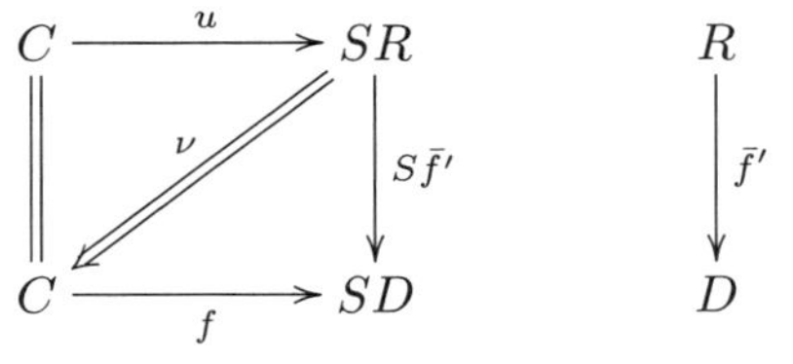

then there exists a unique 2-cell $\nu' : \bar{f}' \Rightarrow f'$ whose $\phi$ image factors $\nu$ via the universal arrow $\mu(f)$, i.e. $\nu'$ is such that

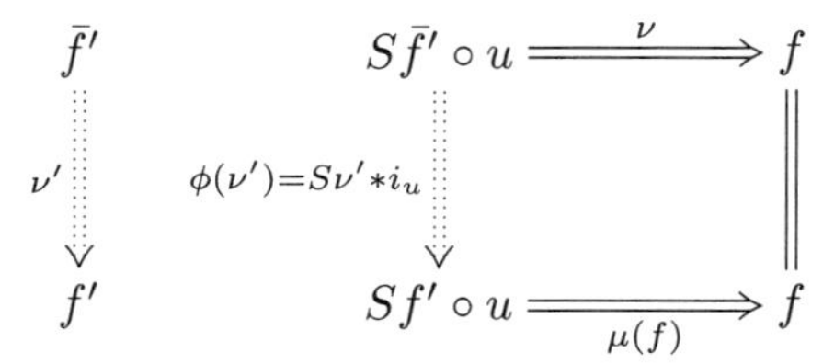

commutes. We also know that $\nu'$ is iso if and only if $\nu$ is iso as in Lemma 9.3. Note that these diagrams are dual to Definition 9.1, although it is the same concept of universal arrow.

One can ask if the equivalences of categories in the definition of biuniversal arrow can be chosen in some natural way as in Remark 3.17. They can in fact as the following theorem shows.

THEOREM 9.5. *Let $u : C \to SR$ be a biuniversal arrow from $C$ to the pseudo functor $S$ as in Definition 9.4. Let $\phi_D : Mor_{\mathcal{D}}(R, D) \to Mor_{\mathcal{C}}(C, SD)$ be the functor defined by $f' \mapsto Sf' \circ u$ and $\gamma \mapsto S\gamma * i_u$. Then $D \mapsto \phi_D$ is a pseudo natural transformation $Mor_{\mathcal{D}}(R, -) \Rightarrow Mor_{\mathcal{C}}(C, S-)$. For $D \in Obj\,\mathcal{D}$ let $\psi_D : Mor_{\mathcal{C}}(C, SD) \to Mor_{\mathcal{D}}(R, D)$ be a right adjoint to $\phi_D$ such that the unit $\eta_D : 1_{Mor_{\mathcal{D}}(R,D)} \Rightarrow \psi_D \circ \phi_D$ and the counit $\varepsilon_D : \phi_D \circ \psi_D \Rightarrow 1_{Mor_{\mathcal{C}}(C,SD)}$ are natural isomorphisms. Then $D \mapsto \psi_D$ is a pseudo natural transformation and $D \mapsto \eta_D$ and $D \mapsto \varepsilon_D$ are iso modifications $i_{Mor_{\mathcal{D}}(R,-)} \leadsto \psi \odot \phi$ and $\phi \odot \psi \leadsto i_{Mor_{\mathcal{C}}(C,S-)}$ which satisfy the triangle identities.*

*Proof:* Let $F, G : \mathcal{D} \to Cat$ be the pseudo functors defined by $F(D) = Mor_{\mathcal{D}}(R, D)$ and $G(D) = Mor_{\mathcal{C}}(C, SD)$. Then $F$ is a strict 2-functor. One can prove that $\phi : F \Rightarrow G$ is a pseudo natural transformation by defining the coherence 2-cell $\tau$ in terms of $\gamma^S$ and then using the unit and composition axioms for $S$ to prove the unit and composition axioms for $\phi$. After doing that, we are in the setup of Lemma 9.9, from which everything else follows. □

In analogy to the uniqueness statement for universal arrows, we have a uniqueness statement for biuniversal arrows. It requires the concept of pseudo isomorphism in a 2-category.

DEFINITION 9.6. Let $\mathcal{D}$ be a 2-category and $f : R \to R'$ a morphism in $\mathcal{D}$. Then $f$ is a *pseudo isomorphism* if there exists a morphism $g : R' \to R$ and iso 2-cells $g \circ f \Rightarrow 1_R$ and $g \circ f \Rightarrow 1_{R'}$. A pseudo isomorphism is also called an *equivalence*.

LEMMA 9.7. *Let $S : \mathcal{D} \to \mathcal{C}$ be a pseudo functor. Let $u_1 : C \to SR_1$ and $u_2 : C \to SR_2$ be biuniversal arrows from $C$ to $S$. Then there exists a pseudo isomorphism $g' : R_1 \to R_2$ in $\mathcal{D}$ and an iso 2-cell as in (9.1).*

(9.1)
$$\begin{array}{ccccc} C & \xrightarrow{u_1} & SR_1 & & R_1 \\ \| & \mu_1(u_2) & \downarrow Sg' & & \downarrow g' \\ C & \xrightarrow{u_2} & SR_2 & & R_2 \end{array}$$

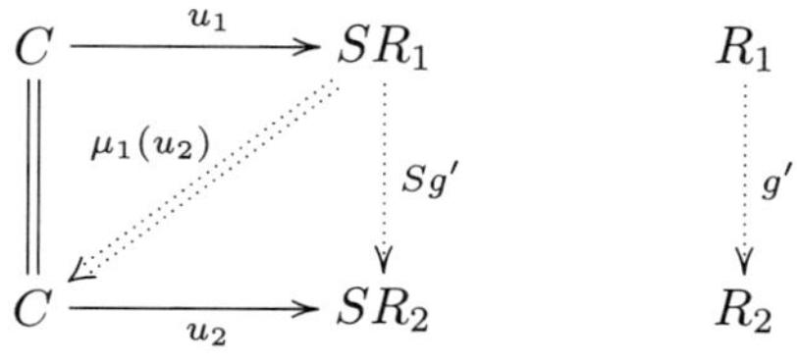

*Moreover, if $\bar{g}'$ and $\nu$ are a morphism and an iso 2-cell that also fill in the diagram, then $\bar{g}'$ and $g'$ are isomorphic via the unique 2-cell $\nu' : \bar{g}' \to g'$ such that $\mu_1(u_2) \circ (S\nu' * i_{u_1}) = \nu$.*

*Proof:* The biuniversality of $u_1$ and $u_2$ guarantees the existence of arrows $f', g'$, and $h'$ and iso 2-cells $\mu_1(u_2), \mu_2(u_1)$, and $\mu_1(u_1)$ to fill in the following diagrams.

(9.2)

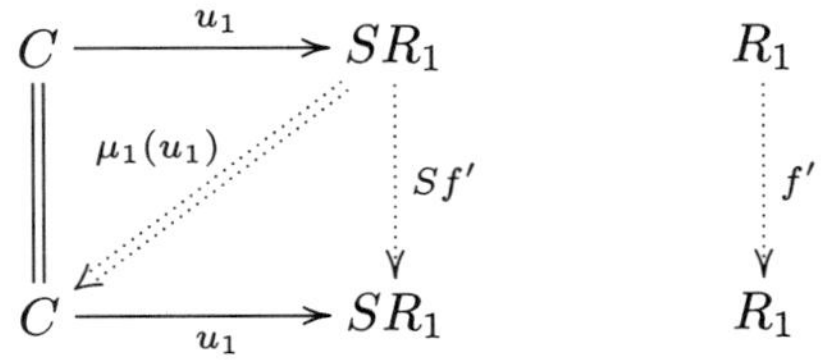


(9.3)

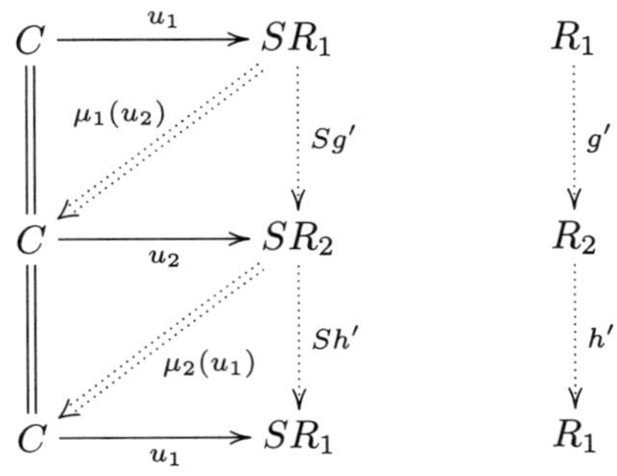


The arrow $1_{R_1}$ also fills in the diagram

(9.4)

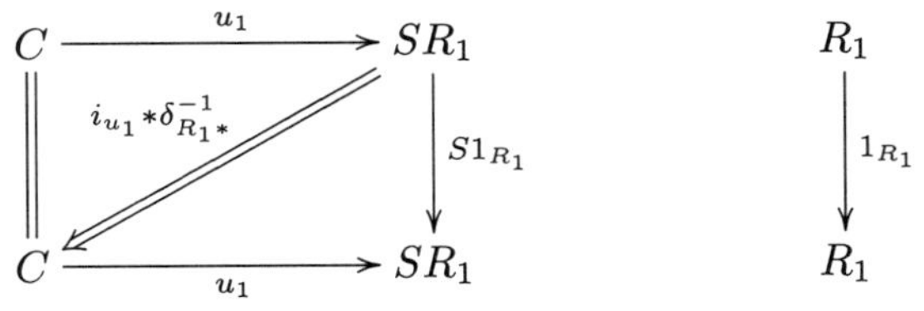


with an iso 2-cell. Diagram (9.3) combined appropriately with $(\gamma^S_{g',h'})^{-1}$ gives an iso 2-cell $h' \circ g' \Rightarrow f'$ by the comments after the definition of biuniversal arrow. Similarly, diagram (9.4) gives an iso 2-cell $1_{R_1} \Rightarrow f'$ for the same reason. Combining these two iso 2-cells appropriately gives an iso 2-cell $h' \circ g' \Rightarrow 1_{R_1}$. By a similar argument we obtain an iso 2-cell $g' \circ h' \Rightarrow 1_{R_2}$. Thus $g' : R_1 \to R_2$ is a pseudo isomorphism. The iso 2-cell between $\bar{g}'$ and $g'$ is also guaranteed by the comments after the definition of biuniversal arrow in 9.4. □

After these preparations involving biuniversal arrows, we can now introduce the main concept of this chapter.

DEFINITION 9.8. Let $\mathcal{X}$ and $\mathcal{A}$ be 2-categories. A *biadjunction* $\langle F, G, \phi \rangle : \mathcal{X} \rightharpoonup \mathcal{A}$ consists of the following data

- Pseudo functors

$$\mathcal{X} \underset{G}{\overset{F}{\rightleftarrows}} \mathcal{A}$$

  between 2-categories

- For all $X \in Obj\ \mathcal{X}$ and all $A \in Obj\ \mathcal{A}$ an equivalence of categories $\phi_{X,A} : Mor_{\mathcal{A}}(FX, A) \to Mor_{\mathcal{X}}(X, GA)$ assigned in such a way to make $\phi$ into a pseudo natural transformation in each variable between the following pseudo functors of two variables.

$$\begin{array}{ccccc} \mathcal{X}^{op} \times \mathcal{A} & \xrightarrow{F^{op} \times 1_{\mathcal{A}}} & \mathcal{A}^{op} \times \mathcal{A} & \xrightarrow{Mor} & Cat \\ & & \Downarrow \phi & & \\ \mathcal{X}^{op} \times \mathcal{A} & \xrightarrow{1_{\mathcal{X}^{op}} \times G} & \mathcal{X}^{op} \times \mathcal{X} & \xrightarrow{Mor} & Cat \end{array}$$

In this situation, $F$ is called a *left biadjoint* for $G$ and $G$ is called a *right biadjoint* for $F$.

Recall again that a *biadjoint* is called a *lax adjoint* in [**25**], [**26**], and [**27**]. The degree of uniqueness of a left biadjoint (if a left biadjoint exists), will be dealt with at the end of this chapter. One can ask whether or not an adjoint functor $\psi_{X,A} : Mor_{\mathcal{X}}(X, GA) \to Mor_{\mathcal{A}}(FX, A)$ to $\phi_{X,A}$ can be chosen in a natural way. This is similar to the question answered in Remark 3.17 for bicolimits. To show that right adjoints can be chosen in a pseudo natural way, we need the following lemma.

LEMMA 9.9. *Let $F, G : \mathcal{A} \to Cat$ be pseudo functors and $F$ a strict 2-functor. Suppose we have a pseudo natural transformation $\phi : F \Rightarrow G$ such that $\phi_A : FA \to GA$ is an equivalence of categories for all $A \in Obj\ \mathcal{A}$. For each $A \in Obj\ \mathcal{A}$, let $\psi_A : GA \to FA$ be a right adjoint to $\phi_A$ such that the unit $\eta_A : 1_{FA} \Rightarrow \psi_A \circ \phi_A$ and counit $\varepsilon_A : \phi_A \circ \psi_A \Rightarrow 1_{GA}$ are natural isomorphisms. Then $A \mapsto \psi_A$ is a pseudo natural transformation $G \Rightarrow F$. The assignments $A \mapsto \eta_A$ and $A \mapsto \varepsilon_A$ define iso modifications $\eta : i_F \rightsquigarrow \psi \odot \phi$ and $\varepsilon : \phi \odot \psi \rightsquigarrow i_G$ respectively. Furthermore, $\eta$ and $\varepsilon$ satisfy the triangle identities.*

*Proof:* For all $A \in Obj\ \mathcal{A}$ there exists such a right adjoint $\psi_A$ because $\phi_A$ is an equivalence of categories.

To show that $A \mapsto \psi_A$ is a pseudo natural transformation, we need to define the coherence 2-cell $\tau'_f$ for each morphism $f$ of $\mathcal{A}$, show that it is natural, it satisfies the unit axiom, and that it satisfies the composition axiom.

For a morphism $f : A \to B$ in $\mathcal{A}$ let $\tau_f : Gf \circ \phi_A \Rightarrow \phi_B \circ Ff$ denote the coherence 2-cell belonging to the pseudo natural transformation $\phi$. Define $\tau'_f$ :

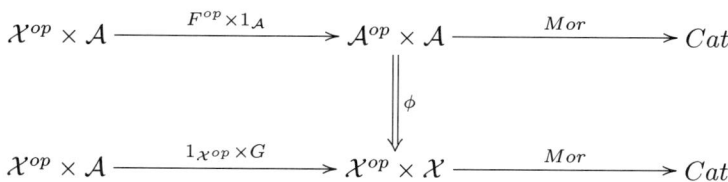

$Ff \circ \psi_A \Rightarrow \psi_B \circ Gf$ to be the composition of the 2-cells in diagram (9.5).

(9.5)

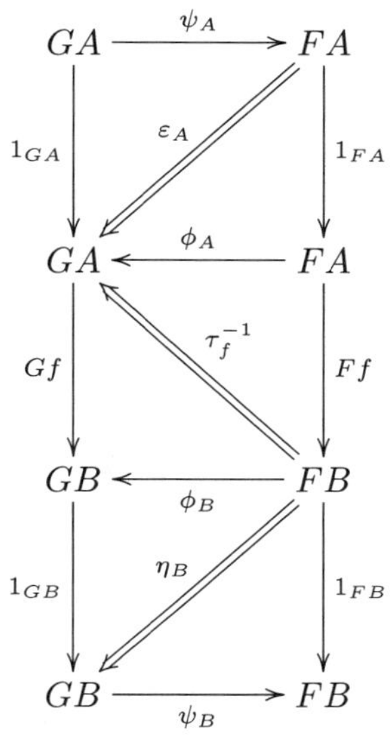


We claim that the assignment $f \mapsto \tau'_f$ is natural in $f$. To see this, let $f, g : A \to B$ be morphisms in $\mathcal{A}$ and $\mu : f \Rightarrow g$ a 2-cell in $\mathcal{A}$. Then $\tau'_f$ is the composition of the top row of 2-cells in diagram (9.6) and $\tau'_g$ is the bottom composition.

(9.6)

$$\begin{array}{ccccccc}
1_{FB}\circ Ff\circ\psi_A & \xRightarrow{\eta_B * i_{Ff} * i_{\psi_A}} & \psi_B\circ\phi_B\circ Ff\circ\psi_A & \xRightarrow{i_{\psi_B}*\tau_f^{-1}*i_{\psi_A}} & \psi_B\circ Gf\circ\phi_A\circ\psi_A & \xRightarrow{i_{\psi_B}*i_{Gf}*\varepsilon_A} & \psi_B\circ Gf\circ 1_{GA}\\
\Downarrow {\scriptstyle i_{1_{FB}}*F\mu*i_{\psi_A}} & & \Downarrow {\scriptstyle i_{\psi_B\circ\phi_B}*F\mu*i_{\psi_A}} & & \Downarrow {\scriptstyle i_{\psi_B}*G\mu*i_{\phi_A\circ\phi_A}} & & \Downarrow {\scriptstyle i_{\psi_B}*G\mu*i_{1_{GA}}}\\
1_{FB}\circ Fg\circ\psi_A & \xRightarrow[\eta_B * i_{Fg} * i_{\psi_A}]{} & \psi_B\circ\phi_B\circ Fg\circ\psi_A & \xRightarrow[i_{\psi_B}*\tau_g^{-1}*i_{\psi_A}]{} & \psi_B\circ Gg\circ\phi_A\circ\psi_A & \xRightarrow[i_{\psi_B}*i_{Gg}*\varepsilon_A]{} & \psi_B\circ Gg\circ 1_{GA}
\end{array}$$

The left square and the right square commute because of the interchange law and the defining property of identity 2-cells. The middle square commutes because $f \mapsto \tau_f$ is natural by the definition of $\phi$ pseudo natural. Hence the outermost rectangle commutes and $f \mapsto \tau'_f$ is natural.

We claim that $\tau'$ satisfies the unit axiom for pseudo natural transformations. Since $F$ is strict, proving the coherence diagram reduces to proving that $\tau'_{1_A} = i_{\psi_A} * \delta^G_{A*}$. Using the definition of $\tau'$ above and the unit axiom for $\tau$ we see that $\tau'_{1_A}$

is the composition of 2-cells in diagram (9.7).

(9.7)

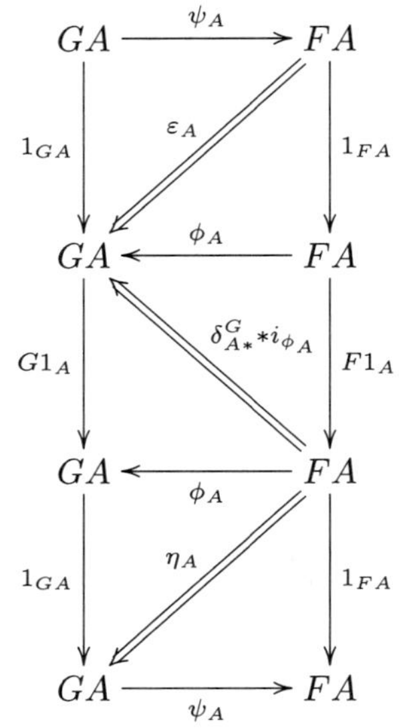


But the composition of 2-cells in (9.7) is the same as the composition of 2-cells in (9.8) by the interchange law.

(9.8)

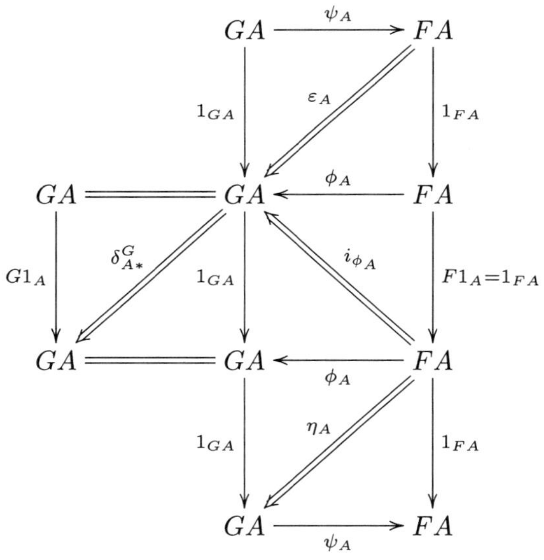


By one of the triangle identities we see that the right three squares of (9.8) collapse to $i_{\psi_A}$ and therefore (9.7) is the same as $i_{\psi_A} * \delta^G_{A*}$. Hence $\tau'_{1_A} = i_{\psi_A} * \delta^G_{A*}$ and the unit axiom is satisfied.

We claim that $\tau'$ satisfies the composition axiom for pseudo natural transformations. Let $A \xrightarrow{f} B \xrightarrow{g} C$ be morphisms in $\mathcal{A}$. Since $F$ is a strict 2-functor, proving the composition coherence reduces to proving that
$\tau'_{g \circ f} = (i_{\psi_C} * \gamma^G_{f,g}) \odot (\tau'_g * i_{Gf}) \odot (i_{Fg} * \tau'_f)$. Following the same approach as for the

unit axiom, we write out $\tau'_{g\circ f}$ in (9.9).

(9.9)

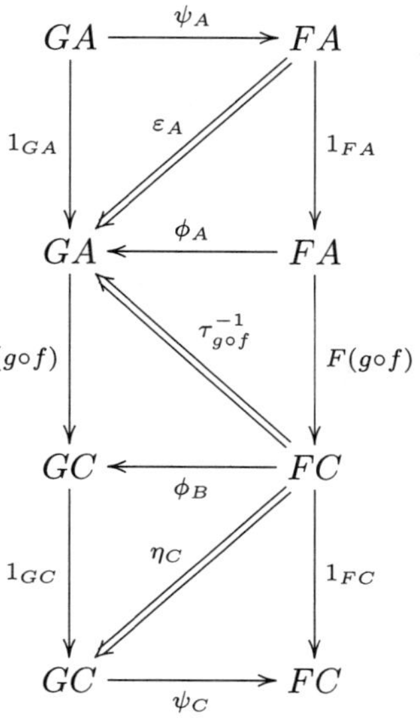


Using the composition axiom for $\tau$ and writing the 2-cells more compactly we see that the composition of 2-cells in diagram (9.9) is the same as in diagram (9.10).

(9.10)

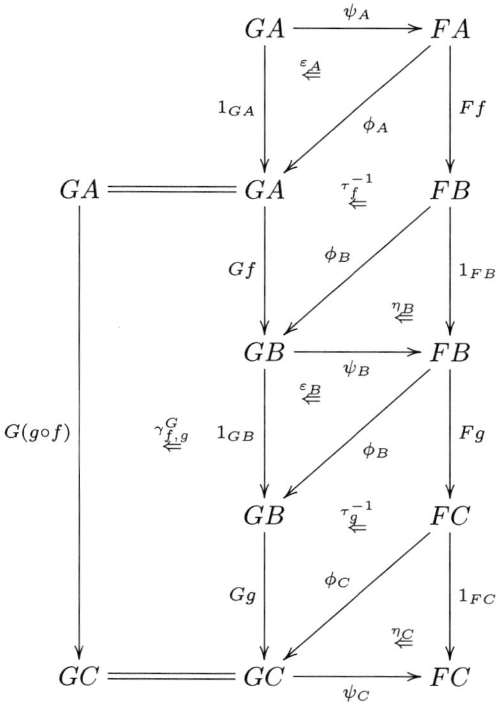


The middle parallelogram involving $\eta_B$ and $\varepsilon_B$ is the same as $i_{\phi_B}$ by the triangle identity. Hence (9.10) is $(i_{\psi_C} * \gamma^G_{f,g}) \odot (\tau'_g * i_{Gf}) \odot (i_{Fg} * \tau'_f)$ and we conclude that $\tau'_{g\circ f} = (i_{\psi_C} * \gamma^G_{f,g}) \odot (\tau'_g * i_{Gf}) \odot (i_{Fg} * \tau'_f)$ as required by the composition axiom.

Thus far we have shown that $A \mapsto \psi_A$ is a pseudo natural transformation $G \Rightarrow F$. Next we show that $A \mapsto \eta_A$ defines a modification $i_F \rightsquigarrow \psi \odot \phi$.

Let $f, g : A \to B$ be morphisms in the 2-category $\mathcal{A}$ and $\gamma : f \Rightarrow g$ a 2-cell. We claim that the compositions in diagrams (3.1) and (3.2) are the same, *i.e.* that $\eta$ is a modification. Our diagrams will of course have $F = G$, $\alpha = i_F$, $\beta = \psi \odot \phi$, and the coherence iso belonging to $i_F$ is trivial while the coherence iso for the composite pseudo natural transformation $\psi \odot \phi$ is $(i_{\psi_B} * \tau_f) \odot (\tau'_f * i_{\phi_A})$ by the remarks on page 13 about coherence isos for a vertical composition of pseudo natural transformations. Then we see that the composition (3.2) is $\eta_B * F\gamma$. We proceed by reducing (3.1) to $\eta_B * F\gamma$. The composition in diagram (3.1) is explicitly (9.11), where we left off the vertical equal signs.

(9.11)

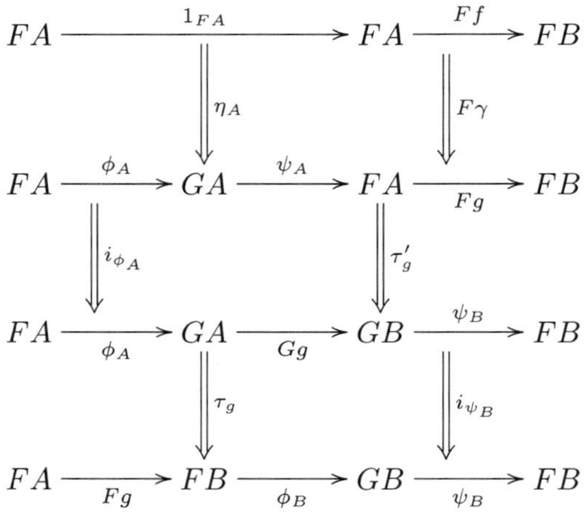


Writing out the definition $\tau'_g$ in (9.11) and including some identities gives (9.12).

(9.12)

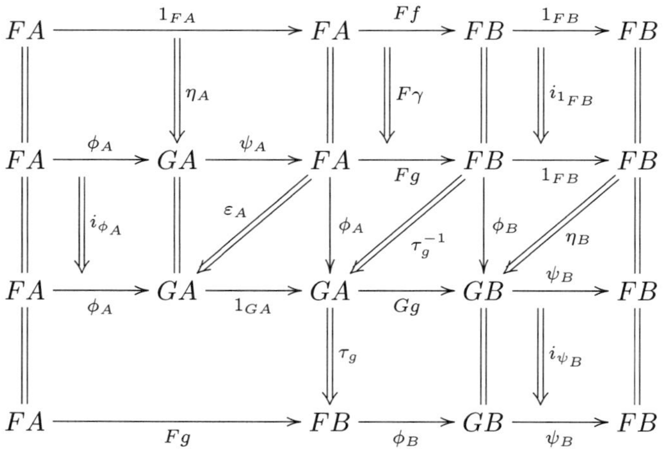


After cancelling $\tau_g$ with $\tau_g^{-1}$ and using one of the triangle identities we see that (9.12) is the same as $\eta_B * F\gamma$. Thus we conclude that (3.1) is the same as (3.2) and that $A \mapsto \eta_A$ is a modification.

One can similarly show that $A \mapsto \varepsilon_A$ is a modification.

The modifications $\eta$ and $\varepsilon$ satisfy the triangle identities because the individual 2-cells $\eta_A$ and $\varepsilon_A$ do. □

Now we use this lemma to prove how the right adjoints $\psi_{X,A} : Mor_{\mathcal{X}}(X, GA) \to Mor_{\mathcal{A}}(FX, A)$ to $\phi_{X,A}$ can be chosen in a pseudo natural way in the following theorem.

THEOREM 9.10. *Let $\langle F, G, \phi \rangle : \mathcal{X} \rightharpoonup \mathcal{A}$ be a biadjunction. For all $X \in Obj\ \mathcal{X}$ and all $A \in Obj\ \mathcal{A}$ let $\psi_{X,A} : Mor_{\mathcal{X}}(X, GA) \to Mor_{\mathcal{A}}(FX, A)$ be a right adjoint to $\phi_{X,A}$ such that the unit $\eta_{X,A} : 1_{Mor_{\mathcal{A}}(FX,A)} \Rightarrow \psi_{X,A} \circ \phi_{X,A}$ and the counit $\varepsilon_{X,A} : \phi_{X,A} \circ \psi_{X,A} \Rightarrow 1_{Mor_{\mathcal{X}}(X,GA)}$ are natural isomorphisms. Then the assignment $(X, A) \mapsto \psi_{X,A}$ is pseudo natural in each variable. Moreover, the assignments $(X, A) \mapsto \eta_{X,A}$ and $(X, A) \mapsto \varepsilon_{X,A}$ comprise modifications in each variable of the form $\eta : i_{Mor_{\mathcal{A}}(F-,-)} \rightsquigarrow \psi \odot \phi$ and $\varepsilon : \phi \odot \psi \rightsquigarrow i_{Mor_{\mathcal{X}}(-,G-)}$.*

*Proof:* We prove the pseudo naturality and modification in the second variable. The first variable is similar. Let $\bar{F}$ respectively $\bar{G}$ be the pseudo functor $\mathcal{A} \to Cat$ obtained by holding $X$ fixed in the first respectively second row in Definition 9.8. See the proof of Lemma 9.15 for a precise description of $\bar{F}$ and $\bar{G}$. The pseudo functor $\bar{F}$ is actually a strict 2-functor because it is the composition of strict 2-functors. If we drop the notation $X$ in all occurrences, we see that we are precisely in the setup of Lemma 9.9. This proves the theorem for the second variable. To prove it for the first variable we only need to prove an analogue of Lemma 9.9 for $F$ pseudo and $G$ strict. □

Next we prove a series of lemmas needed to prove Theorems 9.16 and 9.17.

LEMMA 9.11. *Let $\mathcal{X}$ and $\mathcal{A}$ be 2-categories. Let $\langle F, G, \phi \rangle : \mathcal{X} \rightharpoonup \mathcal{A}$ be a biadjunction and let $\eta_X := \phi_{X,FX}(1_{FX}) : X \to GFX$. Then $\eta_X : X \to G(FX)$ is a biuniversal arrow from $X$ to $G$.*

*Proof:* The assignment $(X, A) \mapsto \phi_{X,A}$ is pseudo natural in each variable by assumption. Let $\tau$ denote the coherence 2-cells for $\phi_{X,-}$. From the definition of pseudo natural transformation $\phi_{X,-}$ we obtain for $f' \in Mor_{\mathcal{A}}(FX, D)$ the following diagram in $Cat$.

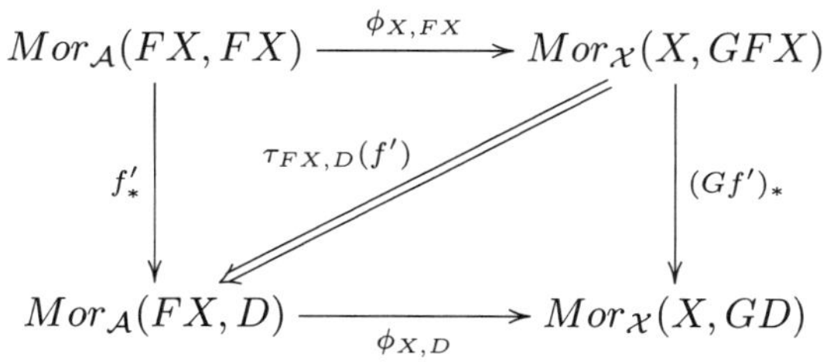


Chasing $1_{FX}$ along this diagram gives a diagram in the 2-category $\mathcal{X}$.

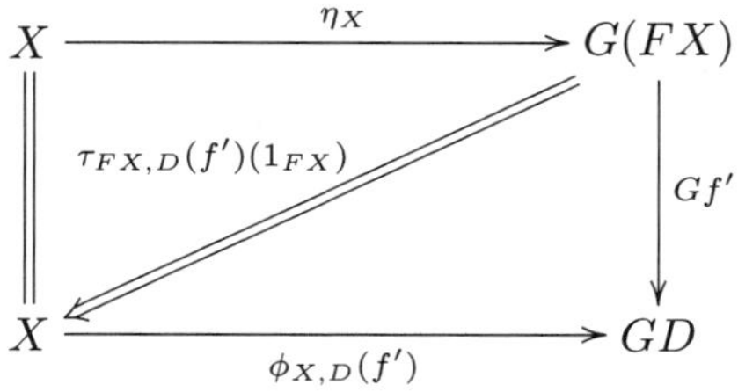


The map $Mor_{\mathcal{A}}(FX, D) \ni f' \mapsto \tau_{FX,D}(f')(1_{FX})$ is natural. This fact combined with the diagram in $\mathcal{X}$ above says that we have a natural isomorphism from the functor $Mor_{\mathcal{A}}(FX, D) \ni f' \mapsto Gf' \circ \eta_X \in Mor_{\mathcal{X}}(X, GD)$ to the functor $f' \mapsto \phi_{X,D}(f')$. From the definition of biadjunction, $\phi_{X,D}$ is an equivalence of categories. Hence $f' \mapsto Gf' \circ \eta_X$ is naturally isomorphic to an equivalence of categories and is therefore itself an equivalence of categories $Mor_{\mathcal{A}}(FX, D) \to Mor_{\mathcal{X}}(X, GD)$. We conclude that $\eta_X$ is a biuniversal arrow. $\square$

LEMMA 9.12. *Let $\mathcal{X}$ and $\mathcal{A}$ be 2-categories. Let $\langle F, G, \phi \rangle : \mathcal{X} \rightharpoonup \mathcal{A}$ be a biadjunction and let $\eta_X := \phi_{X,FX}(1_{FX}) : X \to GFX$. Then the assignment $X \mapsto \eta_X$ is a pseudo natural transformation $1_{\mathcal{X}} \Rightarrow GF$.*

*Proof:* Let $f : X' \to X$ be a morphism in $\mathcal{X}$. Let $\tau$ respectively $\tau'$ denote the coherence 2-cells for the pseudo natural transformation $\phi_{X',-}$ respectively $\phi_{-,FX}$. We must show that we have a 2-cell

$$\begin{array}{ccc} X' & \xrightarrow{\eta_{X'}} & GFX' \\ f \downarrow & & \downarrow GFf \\ X & \xrightarrow{\eta_X} & GFX \end{array}$$

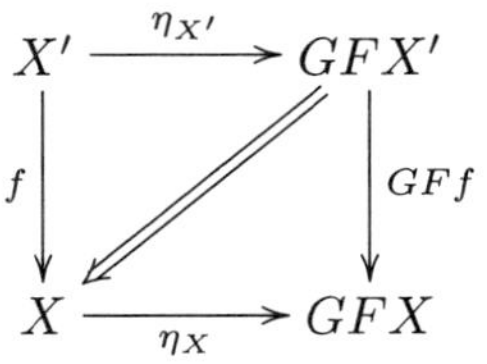

in $\mathcal{X}$ which is natural in $f$ and satisfies the coherences involving $\delta$ and $\gamma$. Since $\phi$ is pseudo natural in each variable we have the diagram

$$\begin{array}{ccccc} Mor_{\mathcal{A}}(FX', FX') & \xrightarrow{(Ff)_*} & Mor_{\mathcal{A}}(FX', FX) & \xleftarrow{(Ff)^*} & Mor_{\mathcal{A}}(FX, FX) \\ \phi_{X',FX'} \downarrow & \overset{\tau_{FX',FX}(Ff)}{\Rightarrow} & \phi_{X',FX} \downarrow & \overset{\tau'_{X,X'}(f^{op})}{\Leftarrow} & \downarrow \phi_{X,FX} \\ Mor_{\mathcal{X}}(X', GFX') & \xrightarrow{(GFf)_*} & Mor_{\mathcal{X}}(X', GFX) & \xleftarrow{f^*} & Mor_{\mathcal{X}}(X, GFX) \end{array}$$

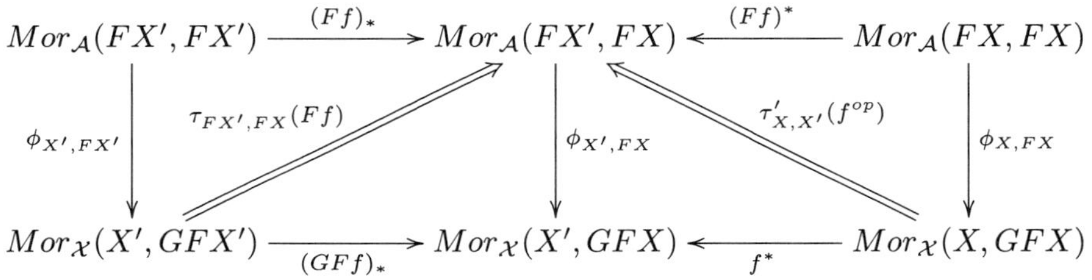

in $Cat$. By chasing $1_{FX'}$ and $1_{FX}$ from the upper corners of this diagram to the center and then down we see that they both get mapped to $\phi_{X',FX}(Ff)$. Chasing the identities in the opposite directions and evaluating the natural transformations at the identities yields a diagram of 2-cells in $\mathcal{X}$.

$$(GFf) \circ \eta_{X'} \xRightarrow{\tau_{FX',FX}(Ff)(1_{FX'})} \phi_{X',FX}(Ff) \xLeftarrow{\tau'_{X,X'}(f^{op})(1_{FX})} \eta_X \circ f$$

These 2-cells are invertible by hypothesis. Let $\tilde{\tau}_{X',X}(f)$ denote the composition from left to right obtained by inverting the second 2-cell. $\tilde{\tau}_{X',X}$ is natural in $f$ because the constituents are natural in $f$. The coherence 2-cells $\tilde{\tau}$ satisfy the coherences with $\delta$ and $\gamma$ from $GF$ also because the individual constituents do. Hence

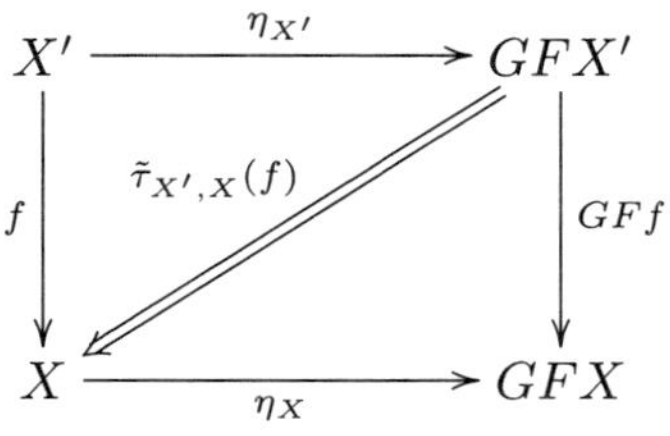

is natural in $f$ and satisfies the required coherences, so $X \mapsto \eta_X$ is a pseudo natural transformation. $\square$

Thus we have seen that given a biadjunction $\phi$ we get a pseudo natural transformation $\eta$ whose arrows are biuniversal arrows. Now we consider the converse of this statement.

LEMMA 9.13. *Let $\mathcal{X}$ and $\mathcal{A}$ be 2-categories. Let $\mathcal{X} \underset{G}{\overset{F}{\rightleftarrows}} \mathcal{A}$ be pseudo functors between 2-categories. Let $\eta : 1_{\mathcal{X}} \Rightarrow GF$ be a pseudo natural transformation such that each arrow $\eta_X : X \to G(FX)$ is a biuniversal arrow from $X$ to $G$. Define $\phi_{X,A}(f) := Gf \circ \eta_X$ for each $f : FX \to A$ and $\phi_{X,A}(\gamma) := G\gamma * i_{\eta_X}$ for each $\gamma : f \Rightarrow f'$. Then $\phi_{X,A} : Mor_{\mathcal{A}}(FX, A) \to Mor_{\mathcal{X}}(X, GA)$ is an equivalence of categories for all $X \in Obj\ \mathcal{X}$ and all $A \in Obj\ \mathcal{A}$.*

*Proof:* The functor $\phi_{X,A}$ is an equivalence since $\eta_X$ is a biuniversal arrow. $\square$

LEMMA 9.14. *Let $\mathcal{X}$ and $\mathcal{A}$ be 2-categories. Let $\mathcal{X} \underset{G}{\overset{F}{\rightleftarrows}} \mathcal{A}$ be pseudo functors between 2-categories. Let $\eta : 1_{\mathcal{X}} \Rightarrow GF$ be a pseudo natural transformation such that each $\eta_X : X \to G(FX)$ is a biuniversal arrow from $X$ to $G$. Let $\phi_{X,A}$ be defined as in Lemma 9.13 above. Then for fixed $A \in Obj\ \mathcal{A}$ the assignment $Obj\ \mathcal{X}^{op} \ni X \mapsto \phi_{X,A}$ denoted $\phi_{-,A}$ is pseudo natural.*

*Proof:* Let $A \in Obj\ \mathcal{A}$ be a fixed object throughout this proof. Let $\bar{F} : \mathcal{X}^{op} \to Cat$ denote the pseudo functor obtained by holding $A$ fixed in the top row in the definition of biadjunction. This means $\bar{F}(X) = Mor_{\mathcal{A}}(FX, A)$, $\bar{F}(f^{op}) = (Ff)^*$, and for $\alpha : f^{op} \Rightarrow (f')^{op}$ in $\mathcal{X}$ the natural transformation $\bar{F}(\alpha) : (Ff)^* \Rightarrow (Ff')^*$ is $h \mapsto i_h * F\alpha$. Note that the morphisms of $\mathcal{X}^{op}$ are formally the opposites of morphisms of $\mathcal{X}$, but the 2-cells of $\mathcal{X}^{op}$ are precisely the same as the 2-cells in $\mathcal{X}$. The vertical composition is the same in both $\mathcal{X}^{op}$ and $\mathcal{X}$, although the horizontal compositions are switched. The pseudo functor $\bar{F}$ is the composition of a pseudo functor and a strict functor. For morphisms $X \xrightarrow{f} Y \xrightarrow{g} Z$ in $\mathcal{X}$ we have $\gamma^{\bar{F}}_{g^{op}, f^{op}} : h \mapsto i_h * \gamma^F_{f,g}$ and for $X \in Obj\ \mathcal{X}^{op}$ we have $\delta^{\bar{F}}_{X*} : h \mapsto i_h * \delta^F_{X*}$ by the rules for composition of pseudo functors. Then $\gamma^{\bar{F}}_{g^{op}, f^{op}} : \bar{F}(f^{op}) \circ \bar{F}(g^{op}) \Rightarrow \bar{F}(f^{op} \circ g^{op})$ and $\delta^{\bar{F}}_{X*} : 1_{\bar{F}X} \Rightarrow \bar{F}(1_X)$. Let $\bar{G}$ denote the strict 2-functor obtained by holding $A$ fixed in the bottom row in the definition of biadjunction. This means $\bar{G}(X) = Mor_{\mathcal{X}}(X, GA)$, $\bar{G}(f^{op}) = f^*$, and for $\alpha : f^{op} \Rightarrow (f')^{op}$ in $\mathcal{X}$ the natural transformation $\bar{G}(\alpha) : \bar{G}(f^{op}) \Rightarrow \bar{G}((f')^{op})$ is the natural transformation $h \mapsto i_h * \alpha$. The 2-functor $\bar{G}$ is the composition of two strict 2-functors and is therefore strict.

In order to prove that $\phi_{-,A}$ is a pseudo natural transformation from $\bar{F}$ to $\bar{G}$ we must display coherence 2-cells $\tau'$ up to which $\phi_{-,A}$ is natural and prove that they satisfy the coherences involving $\delta$ and $\gamma$. Now we describe this $\tau'$ and later prove the coherences. Let $\tilde{\tau}$ denote the coherence 2-cells which make $\eta : 1_{\mathcal{X}} \Rightarrow GF$

pseudo natural, *i.e.* for all $f : X \to Y$ in $\mathcal{X}$ we have

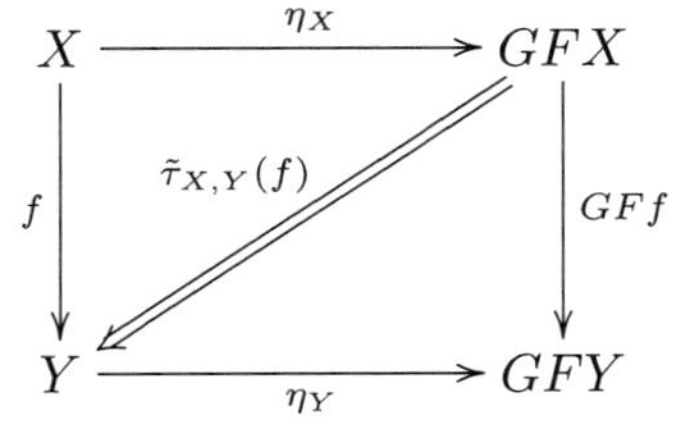


in $\mathcal{X}$. Define a natural isomorphism $\tau'_{f^{op}} = \tau'_{Y,X}(f^{op}) : \bar{G}(f^{op}) \circ \phi_{Y,A} \Rightarrow \phi_{X,A} \circ \bar{F}(f^{op})$ by $h \mapsto (\gamma^G_{Ff,h} * i_{\eta_X}) \odot (i_{Gh} * (\tilde{\tau}_{X,Y}(f))^{-1})$ for $h \in Mor_{\mathcal{A}}(FY, A)$ as in the following diagram.

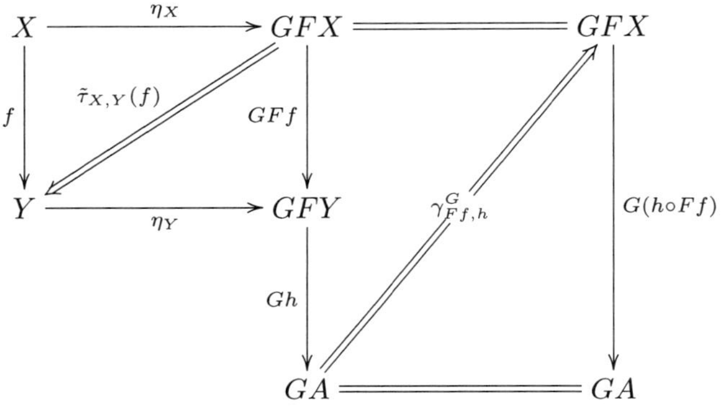


The map $\tau'_{Y,X}(f^{op})$ is a natural transformation because $\gamma^G_{Ff,h}$ is natural in $h$. The assignment $f^{op} \mapsto \tau'_{Y,X}(f^{op})$ is also natural for a similar reason.

We claim that $\tau'$ satisfies the unit axiom for pseudo natural transformations. We must show that the diagram of 2-cells in $Cat$

(9.13)
$$\begin{array}{ccccc}
\phi_{X,A} & = & 1_{\bar{G}X} \circ \phi_{X,A} & = & \bar{G}(1_X) \circ \phi_{X,A} \\
\| & & & & \Downarrow \tau'_{1_X^{op}} \\
\phi_{X,A} \circ 1_{\bar{F}X} & & \xRightarrow{\; i_{\phi_{X,A}} * \delta^{\bar{F}}_{X*} \;} & & \phi_{X,A} \circ \bar{F}(1_X)
\end{array}$$

commutes for all $X \in Obj\, \mathcal{X}$. After we evaluate this diagram on a morphism $h : FX \to A$ of $\mathcal{A}$ we obtain the diagram of 2-cells

(9.14)
$$\begin{array}{ccccc}
Gh \circ \eta_X & = & Gh \circ \eta_X & = & Gh \circ \eta_X \circ 1_X \\
\| & & & & \Downarrow i_{Gh} * (\tilde{\tau}_{X,X}(1_X))^{-1} \\
 & & & & Gh \circ GF1_X \circ \eta_X \\
 & & & & \Downarrow \gamma^G_{F1_X,h} * i_{\eta_X} \\
G(h \circ 1_{FX}) \circ \eta_X & & \xRightarrow{\; G(i_h * \delta^F_{X*}) * i_{\eta_X} \;} & & G(h \circ F(1_X)) \circ \eta_X
\end{array}$$

in $\mathcal{X}$. Since $\eta : 1_{\mathcal{X}} \Rightarrow GF$ is a pseudo natural transformation from the strict 2-functor to the composition $G \circ F$ of pseudo functors, its unit axiom for $\tilde{\tau}$ simplifies to the following commutative diagram.

$$\begin{array}{ccc} 1_{GFX} \circ \eta_X & \xRightarrow{(G(\delta^F_{X*})\odot\delta^G_{FX*})*i_{\eta_X}} & GF1_X \circ \eta_X \\ & \diagdown\!\!\diagdown & \Big\Downarrow \tilde{\tau}_{X,X}(1_X) \\ & & \eta_X \circ 1_X \end{array}$$

Hence $(\tilde{\tau}_{X,X}(1_X))^{-1} = (G(\delta^F_{X*}) \odot \delta^G_{FX*}) * i_{\eta_X}$ as 2-cells. Note also that $\delta^{GF}_{X*} = (G(\delta^F_{X*}) \odot \delta^G_{FX*})$ by the definition of composition of pseudo functors. Using this, we see that diagram (9.14) becomes the outermost rectangle of the following diagram.

$$\begin{array}{ccccc} Gh \circ 1_{GFX} \circ \eta_X & & \xRightarrow{i_{Gh}*\delta^{GF}_{X*}*i_{\eta_X}} & & Gh \circ GF1_X \circ \eta_X \\ & \searrow^{i_{Gh}*\delta^G_{FX*}*i_{\eta_X}} & & \nearrow^{i_{Gh}*G(\delta^F_{X*})*i_{\eta_X}} & \\ \Big\Downarrow i_{Gh}*i_{\eta_X} & & Gh \circ G1_{FX} \circ \eta_X & & \Big\Downarrow \gamma^G_{F1_X,h}*i_{\eta_X} \\ & \swarrow_{\gamma^G_{1_{FX},h}*i_{\eta_X}} & & & \\ G(h \circ 1_{FX}) \circ \eta_X & & \xRightarrow[G(i_h*\delta^F_{X*})*i_{\eta_X}]{} & & G(h \circ F1_X) \circ \eta_X \end{array}$$

The upper left vertex of this diagram is the upper right vertex of diagram (9.14) and the composition of the top arrow and right vertical arrow of this diagram is the right vertical arrow of diagram (9.14). The top triangle of this diagram commutes by definition. The left triangle commutes by the unit axiom of the pseudo functor $G$ applied to the morphism $h : FX \to A$ of $\mathcal{A}$. The right quadrilateral commutes by the naturality of $\gamma^G_{-,h}$ and because $G(i_h * \delta^F_{X*}) = i_{Gh} * G(\delta^F_{X*})$. The morphism $\eta_X$ and the 2-cell $i_{\eta_X}$ just tag along. Hence the outermost rectangle commutes and diagram (9.14) commutes. This implies that diagram (9.13) commutes. We conclude that $\tau'$ satisfies the unit axiom required for $\phi_{-,A}$ to be a pseudo natural transformation.

We claim that $\tau'$ satisfies the composition axiom required for $\phi_{-,A}$ to be a pseudo natural transformation. We must prove for all morphisms $X \xrightarrow{f} Y \xrightarrow{g} Z$ of $\mathcal{X}$, *i.e.* for all morphisms $Z \xrightarrow{g^{op}} Y \xrightarrow{f^{op}} X$ of $\mathcal{X}^{op}$, the diagram of 2-cells in $Cat$ (9.15)

$$\begin{array}{ccccc} \bar{G}(f^{op}) \circ \bar{G}(g^{op}) \circ \phi_{Z,A} & \Longrightarrow & \bar{G}(f^{op}) \circ \phi_{Y,A} \circ \bar{F}(g^{op}) & \Longrightarrow & \phi_{X,A} \circ \bar{F}(f^{op}) \circ \bar{F}(g^{op}) \\ \Big\Downarrow & & & & \Big\Downarrow \\ \bar{G}(f^{op} \circ g^{op}) \circ \phi_{Z,A} & & \Longrightarrow & & \phi_{X,A} \circ \bar{F}(f^{op} \circ g^{op}) \end{array}$$

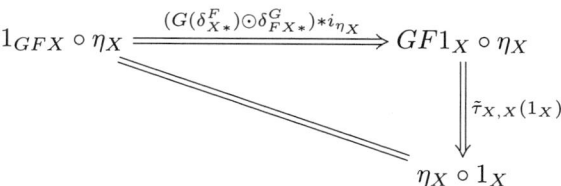

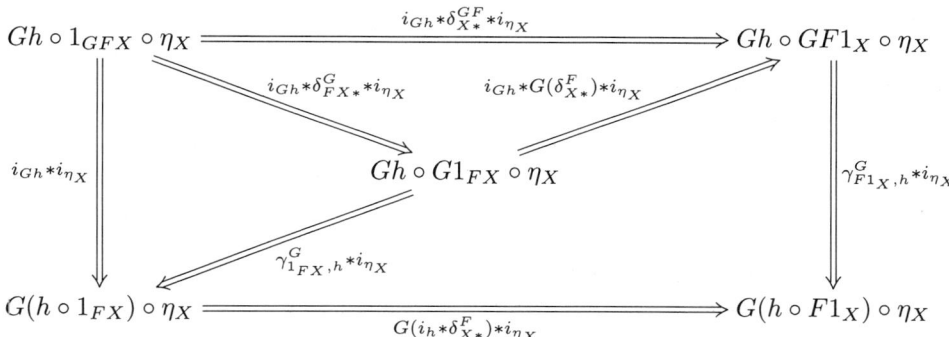

commutes. More precisely the diagram of 2-cells in $Cat$
(9.16)

$$\begin{array}{ccccc}
f^* \circ g^* \circ \phi_{Z,A} & \xRightarrow{i_{f^*} * \tau'_{g^{op}}} & f^* \circ \phi_{Y,A} \circ (Fg)^* & \xRightarrow{\tau'_{f^{op}} * i_{(Fg)^*}} & \phi_{X,A} \circ (Ff)^* \circ (Fg)^* \\
\Vert & & & & \Big\Downarrow {\scriptstyle i_{\phi_{X,A}} * \gamma^{\bar{F}}_{g^{op},f^{op}}} \\
(g \circ f)^* \circ \phi_{Z,A} & & \xRightarrow[\tau'_{f^{op} \circ g^{op}}]{} & & \phi_{X,A} \circ (F(g \circ f))^*
\end{array}$$

must commute. We evaluate this diagram on a morphism $h : FZ \to A$ of $\mathcal{A}$, fill in the diagram with more vertices, and cut the result down the middle column to get the left respectively right half on page 97. These are diagrams of 2-cells in $\mathcal{X}$. Subdiagram (I) commutes by the composition axiom applied to the morphisms $X \xrightarrow{f} Y \xrightarrow{g} Z$ for the pseudo natural transformation $\eta : 1_{\mathcal{X}} \Rightarrow GF$ with its coherence 2-cells $\tilde{\tau}$. Subdiagram (II) commutes by the composition axiom applied to the morphisms $Ff, Fg, h$ for the pseudo functor $G$ with its coherence 2-cells $\gamma^G$. The fifth arrow which is an equality symbol was only drawn for convenience. Subdiagram (III) commutes by the naturality of $\gamma^G$. All other subdiagrams commute by definition or by the interchange law. Therefore the outermost rectangle commutes when we put the two halves together. This outermost rectangle is diagram (9.16) evaluated on the morphism $h : FZ \to A$ of $\mathcal{A}$. Hence (9.16) and (9.15) commute. We conclude that $\tau'$ satisfies the composition axiom required for $\phi_{-,A}$ to be a pseudo natural transformation.

Since $\phi_{-,A}$ with coherence 2-cells $\tau'$ satisfies the unit axiom and composition axiom for pseudo natural transformations we conclude that $\phi_{-,A}$ is a pseudo natural transformation for fixed $A \in Obj\ \mathcal{A}$.

□

LEMMA 9.15. *Let $\mathcal{X}$ and $\mathcal{A}$ be 2-categories. Let $\mathcal{X} \underset{G}{\overset{F}{\rightleftarrows}} \mathcal{A}$ be pseudo functors between 2-categories. Let $\eta : 1_{\mathcal{X}} \Rightarrow GF$ be a pseudo natural transformation such that each $\eta_X : X \to G(FX)$ is a biuniversal arrow from $X$ to $G$. Let $\phi_{X,A}$ be defined as in Lemma 9.13 above. Then for fixed $X \in Obj\ \mathcal{X}$ the assignment $Obj\ \mathcal{A} \ni A \mapsto \phi_{X,A}$ denoted $\phi_{X,-}$ is pseudo natural.*

*Proof:* Let $X$ be a fixed object of the 2-category $\mathcal{X}$ throughout the proof. We introduce new pseudo functors $\bar{F}$ and $\bar{G}$ different from those in the previous proof. Let $\bar{F} : \mathcal{A} \to Cat$ be the strict 2-functor obtained by fixing $X$ in the top row in the definition of biadjunction. This means $\bar{F}(A) = Mor_{\mathcal{A}}(FX, A)$, $\bar{F}(f) = f_*$, and for $\alpha : f \Rightarrow f'$ we have $\bar{F}(\alpha)$ is the natural transformation $e \mapsto \alpha * i_e$. The 2-functor $\bar{F}$ is strict because it is the composition of two strict 2-functors. Similarly let $\bar{G} : \mathcal{A} \to Cat$ be the pseudo functor obtained by fixing $X$ in the bottom row of the definition of biadjunction. This means $\bar{G}(A) = Mor_{\mathcal{X}}(X, GA)$, $\bar{G}(f) = (Gf)_*$, and for $\alpha : f \Rightarrow f'$ we have $\bar{G}(\alpha)$ is the natural transformation $e \mapsto G(\alpha) * i_e$. The pseudo functor $\bar{G}$ is pseudo because it is the composition of a pseudo functor and a strict functor. The definition of composition of pseudo functors then says that the coherence 2-cells for $\bar{G}$ are $\gamma^{\bar{G}}_{f,g} : e \mapsto \gamma^G_{f,g} * i_e$ for morphisms $f, g$ of $\mathcal{A}$ such that $g \circ f$ exists and $\delta^{\bar{G}}_{A*} : e \mapsto \delta^G_{A*} * i_e$ for $A \in Obj\ \mathcal{A}$. These are natural transformations, *i.e.*

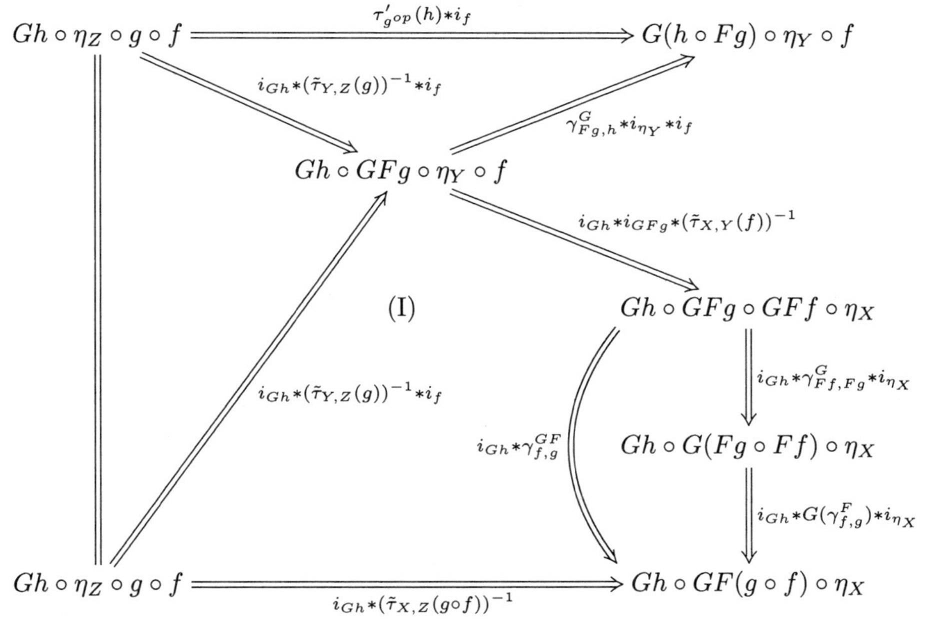


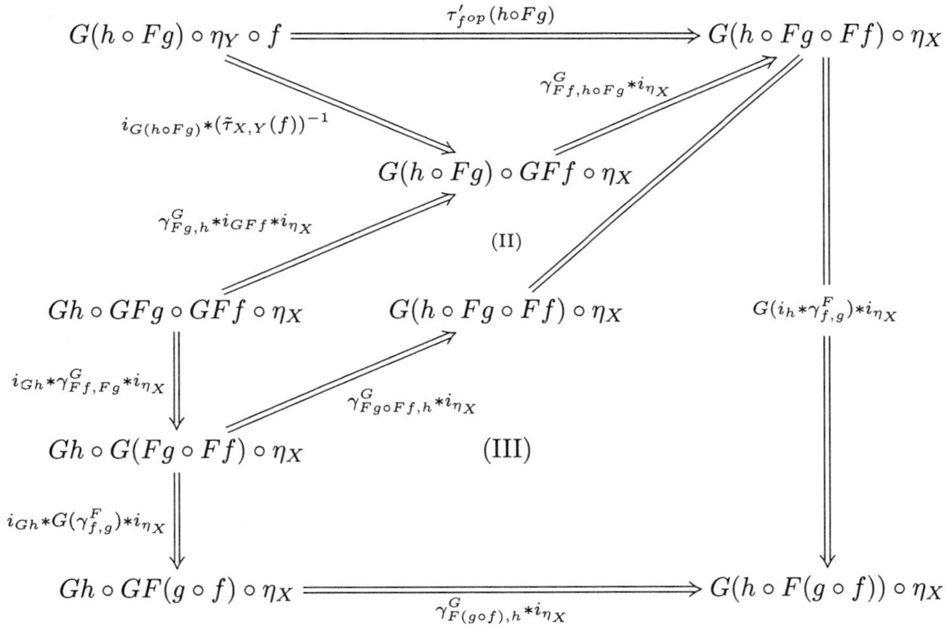


2-cells in $Cat$, such that $\gamma^{\bar{G}}_{f,g} : \bar{G}(g) \circ \bar{G}(f) \Rightarrow \bar{G}(g \circ f)$ and $\delta^{\bar{G}}_{A*} : 1_{\bar{G}(A)} \Rightarrow \bar{G}(1_A)$. They are natural in $f$ and $g$ and they satisfy the required coherences for a pseudo functor.

We must show that $\phi_{X,-}$ is a pseudo natural transformation from $\bar{F}$ to $\bar{G}$. In other words we must display coherence 2-cells $\tau$ up to which $\phi_{X,-}$ is natural and satisfy the coherence diagrams involving $\gamma$ and $\delta$ from $\bar{F}$ and $\bar{G}$. For morphisms

$k : A \to A'$ of $\mathcal{A}$ define $\tau_{A,A'}(k) : e \mapsto \gamma^G_{e,k} * i_{\eta_X}$ to fill in the diagram

$$\begin{array}{ccc} Mor_{\mathcal{A}}(FX, A) & \xrightarrow{\phi_{X,A}} & Mor_{\mathcal{X}}(X, GA) \\ {\scriptstyle k_*}\downarrow & \overset{\tau_{A,A'}(k)}{\swarrow\!\!\!\!\!\!\!\Leftarrow} & \downarrow{\scriptstyle (Gk)_*} \\ Mor_{\mathcal{A}}(FX, A') & \xrightarrow[\phi_{X,A'}]{} & Mor_{\mathcal{A}}(X, GA') \end{array}$$

whose vertices are $\bar{F}(A), \bar{G}(A), \bar{G}(A')$, and $\bar{F}(A')$ read clockwise. The map $\tau_{A,A'}(k)$ is a natural transformation (2-cell in $Cat$) between the indicated functors because $\gamma^G_{e,k}$ is natural in $e$. The assignment $Mor_{\mathcal{A}}(A, A') \ni k \mapsto \tau_{A,A'}(k)$ is a natural transformation $(\circ\phi_{X,A}) \circ \bar{G} \Rightarrow (\phi_{X,A'}\circ) \circ \bar{F}$ because $\gamma^G_{e,k}$ is natural in $k$. Hence this family $\tau$ of natural transformations provides us with a candidate for the coherence 2-cells to make $\phi_{X,-}$ into a pseudo natural transformation.

We claim that $\tau$ satisfies the unit axiom for pseudo natural transformations. This requires a proof that the diagram of 2-cells in $Cat$

$$\begin{array}{ccccc} \phi_{X,A} & \xRightarrow{i_{\phi_{X,A}}} & 1_{\bar{G}A} \circ \phi_{X,A} & \xRightarrow{\delta^{\bar{G}}_{A*} * i_{\phi_{X,A}}} & \bar{G}(1_A) \circ \phi_{X,A} \\ \| & & & & \Downarrow{\scriptstyle \tau_{1_A}} \\ \phi_{X,A} \circ 1_{\bar{F}A} & & = & & \phi_{X,A} \circ \bar{F}(1_A) \end{array}$$

commutes for all $A \in Obj\ \mathcal{A}$. Evaluating this diagram on a morphism $e : FX \to A$ of $\mathcal{A}$ results in the diagram of 2-cells

$$\begin{array}{ccccc} Ge \circ \eta_X & = & 1_{GA} \circ Ge \circ \eta_X & \xRightarrow{\delta^G_{A*} * i_{Ge} * i_{\eta_X}} & G(1_A) \circ Ge \circ \eta_X \\ & \searrow\!\!\!\!\!= & & & \Downarrow{\scriptstyle \gamma^G_{e,1_A} * i_{\eta_X}} \\ & & & & G(1_A \circ e) \circ \eta_X \end{array}$$

in $\mathcal{X}$ which commutes because of the unit axiom for the pseudo functor $G$. Hence $\tau$ satisfies the unit axiom for pseudo natural transformations.

We claim that $\tau$ satisfies the composition axiom for pseudo natural transformations. This requires us to prove for all morphisms $A \xrightarrow{f} B \xrightarrow{g} C$ in $\mathcal{A}$ that the diagram of 2-cells in $Cat$

$$\begin{array}{ccccc} \bar{G}g \circ \bar{G}f \circ \phi_{X,A} & \xRightarrow{i_{\bar{G}g} * \tau_f} & \bar{G}g \circ \phi_{X,B} \circ \bar{F}f & \xRightarrow{\tau_g * i_{\bar{F}f}} & \phi_{X,C} \circ \bar{F}g \circ \bar{F}f \\ {\scriptstyle \gamma^{\bar{G}}_{f,g} * i_{\phi_{X,A}}}\Downarrow & & & & \| \\ \bar{G}(g \circ f) \circ \phi_{X,A} & & \xRightarrow[\tau_{g\circ f}]{} & & \phi_{X,C} \circ \bar{F}(g \circ f) \end{array}$$

commutes. Evaluating this diagram on a morphism $e : FX \to A$ of $\mathcal{A}$ results in the diagram of 2-cells

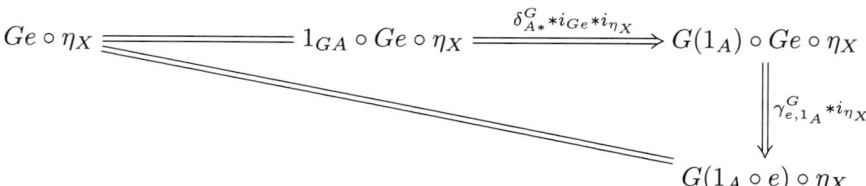

$$Gg \circ Gf \circ Ge \circ \eta_X \xrightarrow{i_{Gg}*\gamma^G_{e,f}*i_{\eta_X}} Gg \circ G(f \circ e) \circ \eta_X \xrightarrow{\gamma^G_{foe,g}*i_{\eta_X}} G(g \circ f \circ e) \circ \eta_X$$

$$\left\Vert\gamma^G_{f,g}*i_{Ge}*i_{\eta_X}\right. \qquad\qquad\qquad\qquad\qquad\qquad\qquad\qquad \left\Vert\right.$$

$$G(g \circ f) \circ G(e) \circ \eta_X \xrightarrow{\gamma^G_{e,g\circ f}*i_{\eta_x}} G(g \circ f \circ e) \circ \eta_X$$

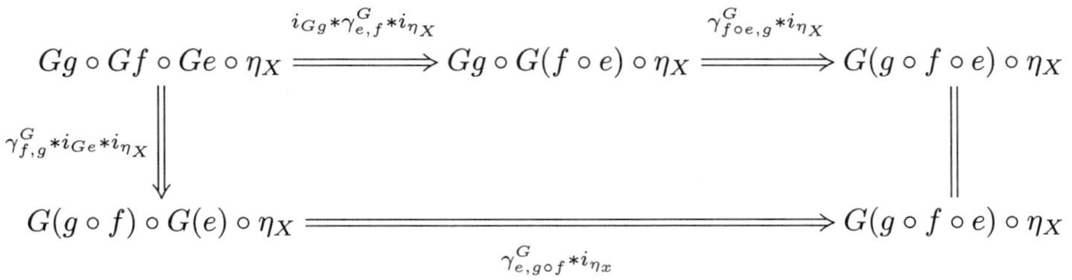

in $\mathcal{X}$, which commutes by the composition axiom for the pseudo functor $G$ applied to $FX \xrightarrow{e} A \xrightarrow{f} B \xrightarrow{g} C$. Hence $\tau$ satisfies the composition axiom for pseudo natural transformations.

We conclude that $\phi_{X,-}$ is a pseudo natural transformation from $\bar{F}$ to $\bar{G}$ with coherence 2-cells defined by $\tau$. □

Now we can finally state and prove the two main theorems of this chapter.

THEOREM 9.16. *Let $\mathcal{X}$ and $\mathcal{A}$ be 2-categories. Let $\mathcal{X} \xrightleftharpoons[G]{F} \mathcal{A}$ be pseudo functors. Then $F$ is a left biadjoint for $G$ if and only if there exists a pseudo natural transformation $\eta : 1_\mathcal{X} \Rightarrow GF$ such that $\eta_X : X \to G(FX)$ is a biuniversal arrow for all $X \in Obj\, \mathcal{X}$.*

*Proof:* This follows immediately from the previous lemmas. □

THEOREM 9.17. *Let $\mathcal{X}$ and $\mathcal{A}$ be 2-categories. Let $\mathcal{X} \xleftarrow{G} \mathcal{A}$ be a pseudo functor. Then there exists a left biadjoint for $G$ if and only if for every object $X \in Obj\, \mathcal{X}$ there exists an object $R \in Obj\, \mathcal{A}$ and a biuniversal arrow $\eta_X : X \to G(R)$ from $X$ to $G$.*

*Proof:* By Lemma 9.11, the existence of a left biadjoint implies the existence of such a biuniversal arrow. Now we prove the other direction. Suppose we have such a biuniversal arrow for each $X \in Obj\, \mathcal{X}$. Define $FX := R$. The object $R \in Obj\, \mathcal{A}$ of course depends on $X$. For $X \in Obj\, \mathcal{X}$ and $A \in Obj\, \mathcal{A}$ let $\phi_{X,A} : Mor_\mathcal{A}(FX, A) \to Mor_\mathcal{X}(X, GA)$ denote the functor $f' \mapsto Gf' \circ \eta_X$ and $\alpha \mapsto G\alpha * i_{\eta_X}$. Let $\psi_{X,A} : Mor_\mathcal{X}(X, GA) \to Mor_\mathcal{A}(FX, A)$ denote a right adjoint equivalence, which exists because $\eta_X$ is a biuniversal arrow. Let $\mu_{X,A} : \phi_{X,A} \circ \psi_{X,A} \Rightarrow 1_{Mor_\mathcal{X}(X,GA)}$ denote a counit for these adjoint functors. All of this implies that for any morphism $f : X \to GA$ there exists a morphism $f' := \psi_{X,A}(f)$ and a 2-cell $\mu_{X,A}(f)$ as in the diagram.

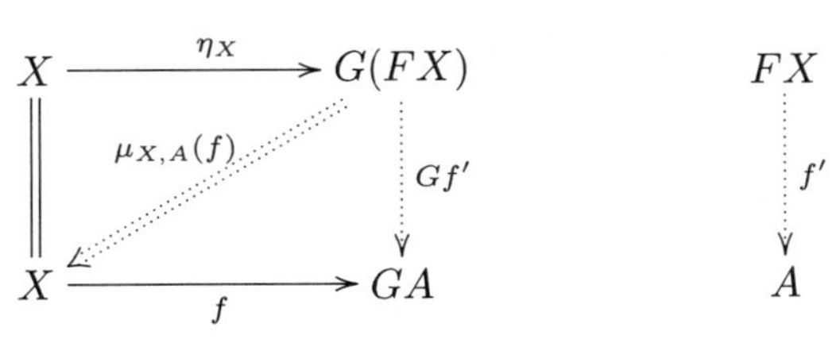

Moreover, this 2-cell $\mu_{X,A}(f)$ is a universal arrow from the functor $\phi_{X,A} \circ \psi_{X,A}$ to the object $f$ because all of the arrows of the counit of an adjunction are universal.

This means that for any other morphism $\bar{f}' : FX \to A$ and 2-cell $\nu$ as in the diagram

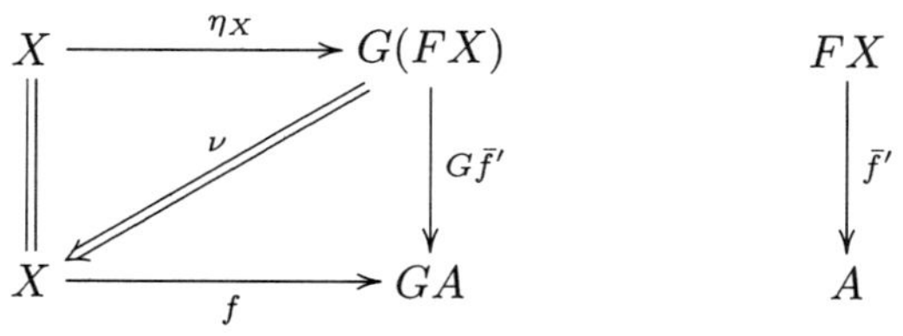


there exists a unique 2-cell $\nu' : \bar{f}' \Rightarrow f'$ such that the following diagram commutes.

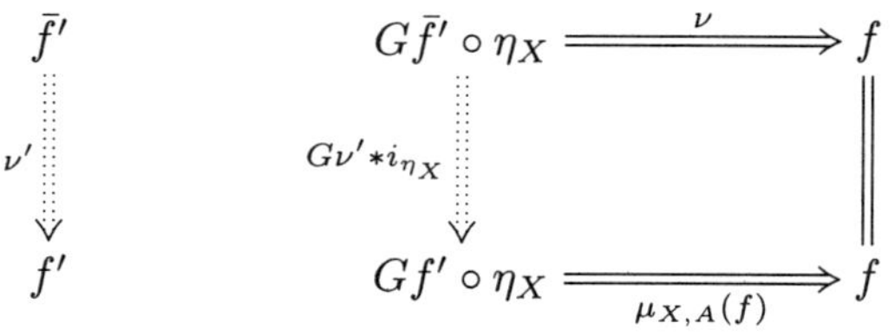


If $\nu$ is iso, this 2-cell $\nu' : \bar{f}' \Rightarrow f'$ is also iso by the comments after Definition 9.4. The uniqueness and iso property of $\nu'$ will be integral to defining the coherence isomorphisms and proving the coherence diagrams below.

After setting up this notation, we define a left biadjoint candidate $F$ for $G$. We already have $F$ defined for objects $X \in Obj\ \mathcal{X}$ above. For any morphism $h : X \to Y$ in $\mathcal{X}$ define $Fh := \psi_{X,FY}(\eta_Y \circ h)$. For morphisms $h, h' : X \to Y$ and any 2-cell $\alpha : h \Rightarrow h'$ in $\mathcal{X}$ define $F\alpha := \psi_{X,FY}(i_{\eta_Y} * \alpha)$. Then the assignment is obviously a functor on any fixed hom category because of the interchange law and because $\psi_{X,FY}$ preserves identity 2-cells and compositions of 2-cells. To define the coherence 2-cells $\delta^F_X$ we now use the uniqueness described above. Note that $F1_X = \psi_{X,FX}(\eta_X \circ 1_X)$ satisfies the diagram

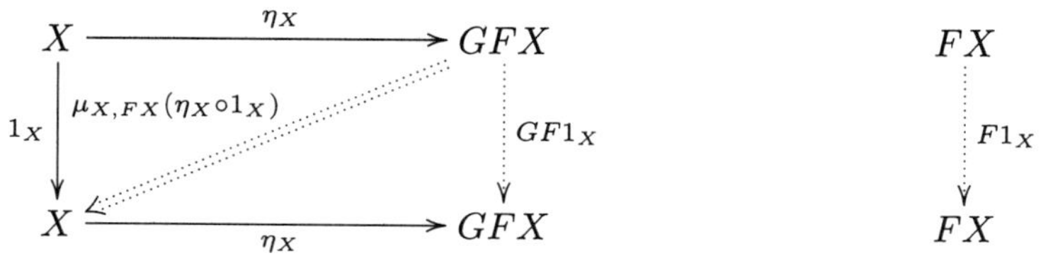


where $\mu_{X,FX}(\eta_X \circ 1_X)$ is universal. The arrow $1_{FX}$ satisfies

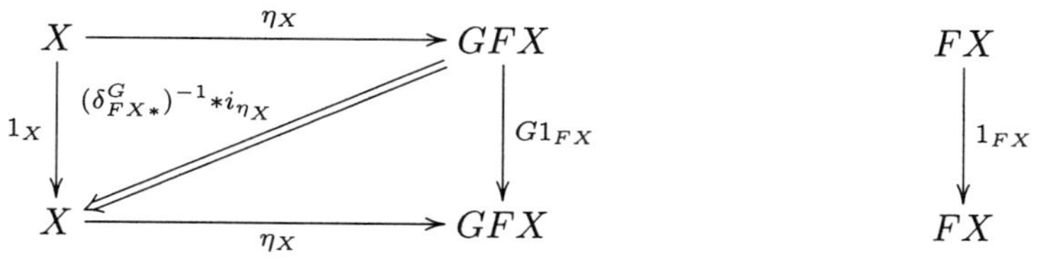

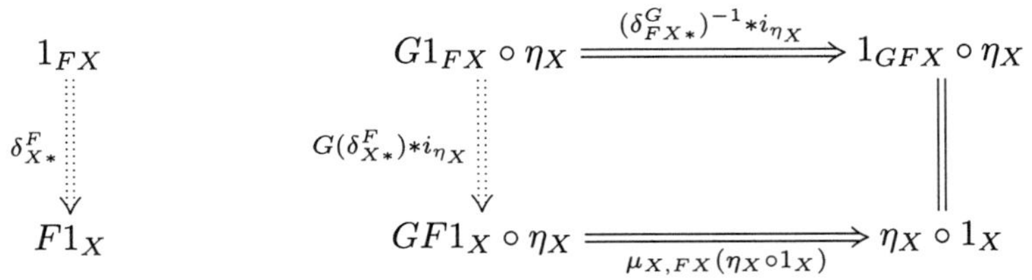

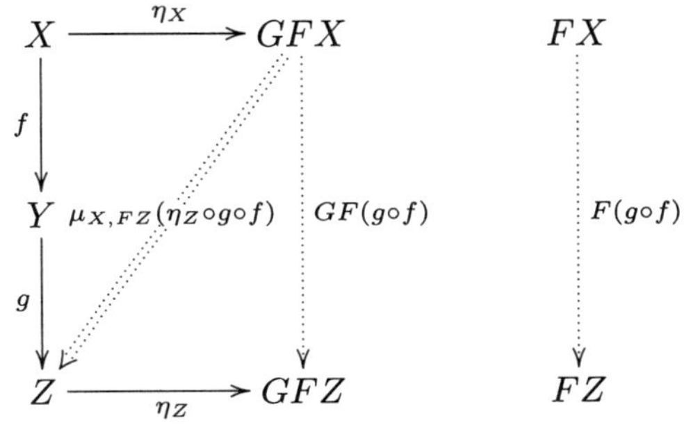

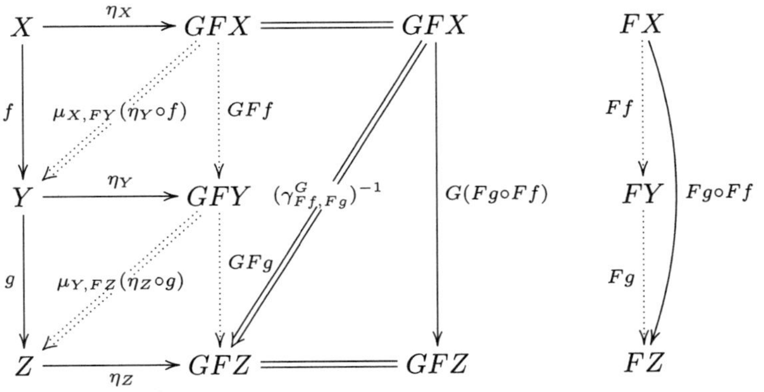

since $G$ is a pseudo functor. Let $\delta^F_{X*} : 1_{FX} \Rightarrow F1_X$ be the unique 2-cell whose $\phi_{X,FX}$ image factors $(\delta^G_{FX*})^{-1} * i_{\eta_X}$.

(9.17)
$$\begin{CD}
1_{FX} @. G1_{FX} \circ \eta_X @>(\delta^G_{FX*})^{-1} * i_{\eta_X}>> 1_{GFX} \circ \eta_X \\
@V\delta^F_{X*}VV @VG(\delta^F_{X*})*i_{\eta_X}VV @| \\
F1_X @. GF1_X \circ \eta_X @>>\mu_{X,FX}(\eta_X \circ 1_X)> \eta_X \circ 1_X
\end{CD}$$

It exists by the universality of $\mu_{X,FX}(\eta_X \circ 1_X)$. The 2-cell $\delta^F_{X*} : 1_{FX} \Rightarrow F1_X$ is iso because $(\delta^G_{FX*})^{-1} * i_{\eta_X}$ is iso. To define $\gamma^F_{f,g}$ for $X \xrightarrow{f} Y \xrightarrow{g} Z$ in $\mathcal{X}$ we similarly use the uniqueness. Note that $F(g \circ f) = \psi_{X,FZ}(\eta_Z \circ g \circ f)$ satisfies the diagram

where the 2-cell $\mu_{X,FZ}(\eta_Z \circ g \circ f)$ is universal. The arrow $Fg \circ Ff$ satisfies

(9.18)
since $G$ is a pseudo functor. Let $\gamma^F_{f,g} : Fg \circ Ff \Rightarrow F(g \circ f)$ be the unique 2-cell whose $\phi_{X,FZ}$ image factors the composition of the 2-cells in (9.18) as follows.

(9.19)
$$\begin{CD}
Fg \circ Ff @. G(Fg \circ Ff) \circ \eta_X @>>> \eta_Z \circ g \circ f \\
@V\gamma^F_{f,g}VV @VG(\gamma^F_{f,g})*i_{\eta_X}VV @| \\
F(g \circ f) @. GF(g \circ f) \circ \eta_X @>>\mu_{X,FZ}(\eta_Z \circ g \circ f)> \eta_Z \circ g \circ f
\end{CD}$$

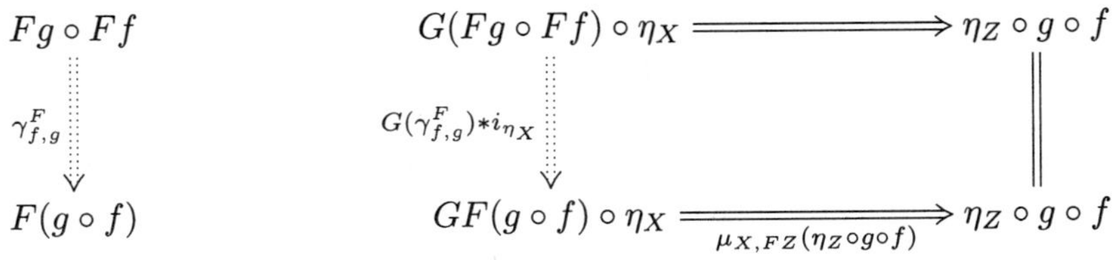

The top horizontal 2-cell in the previous diagram is the composition of the 2-cells in (9.18). The 2-cell $\gamma^F_{f,g} : Fg \circ Ff \Rightarrow F(g \circ f)$ is iso because the composition of

2-cells in (9.18) is iso. Thus we have completely defined a left biadjoint candidate $F$ for $G$. Now we must show that the 2-cells do what they should in order for $F$ to be a pseudo functor.

We claim that $\gamma^F$ is natural in its two variables. We must show for morphisms $X \xrightarrow{f_i} Y \xrightarrow{g_i} Z$ in $\mathcal{X}$ and 2-cells $\alpha : f_1 \Rightarrow f_2$ and $\beta : g_1 \Rightarrow g_2$ in $\mathcal{X}$ that

(9.20)
$$\begin{array}{ccc}
Fg_1 \circ Ff_1 & \xRightarrow{\gamma^F_{f_1,g_1}} & F(g_1 \circ f_1) \\
{\scriptstyle F\beta * F\alpha}\Big\Downarrow & & \Big\Downarrow{\scriptstyle F(\beta*\alpha)} \\
Fg_2 \circ Ff_2 & \xRightarrow[\gamma^F_{f_2,g_2}]{} & F(g_2 \circ f_2)
\end{array}$$

commutes.

Toward this end, consider diagrams (9.21) and (9.22).

(9.21)
$$\begin{array}{ccc}
G(Fg_1 \circ Ff_1) \circ \eta_X & \xRightarrow{\sigma_1} & \eta_Z \circ g_1 \circ f_1 \\
{\scriptstyle G(\gamma^F_{f_1,g_1}) * i_{\eta_X}}\Big\Downarrow & & \Big\| \\
G(F(g_1 \circ f_1)) \circ \eta_X & \xRightarrow{\mu_{X,FZ}(\eta_Z \circ g_1 \circ f_1)} & \eta_Z \circ g_1 \circ f_1 \\
{\scriptstyle GF(\beta*\alpha) * i_{\eta_X}}\Big\Downarrow & & \Big\Downarrow{\scriptstyle i_{\eta_Z} * \beta * \alpha} \\
GF(g_2 \circ f_2) \circ \eta_X & \xRightarrow[\mu_{X,FZ}(\eta_Z \circ g_2 \circ f_2)]{} & \eta_Z \circ g_2 \circ f_2
\end{array}$$

(9.22)
$$\begin{array}{ccc}
G(Fg_1 \circ Ff_1) \circ \eta_X & \xRightarrow{\sigma_1} & \eta_Z \circ g_1 \circ f_1 \\
{\scriptstyle G(F\beta * F\alpha) * i_{\eta_X}}\Big\Downarrow & & \Big\Downarrow{\scriptstyle i_{\eta_Z} * \beta * \alpha} \\
G(Fg_2 \circ Ff_2) \circ \eta_X & \xRightarrow{\sigma_2} & \eta_Z \circ g_2 \circ f_2 \\
{\scriptstyle G(\gamma^F_{f_2,g_2}) * i_{\eta_x}}\Big\downarrow & & \Big\| \\
G(F(g_2 \circ f_2)) \circ \eta_X & \xRightarrow[\mu_{X,FZ}(\eta_Z \circ g_2 \circ f_2)]{} & \eta_Z \circ g_2 \circ f_2
\end{array}$$

The top horizontal 2-cell $\sigma_1$ in both diagrams is the composition of the 2-cells in diagram (9.18) with $f, g$ replaced by $f_1, g_1$ respectively. The bottom horizontal 2-cell in each diagram is $\mu_{X,FZ}(\eta_Z \circ g_2 \circ f_2)$. The center horizontal 2-cell $\sigma_2$ in (9.22) is the composition of the 2-cells in (9.18) with $f, g$ replaced by $f_2, g_2$ respectively. The top rectangle in (9.21) commutes because it is the analogue of (9.19) for $f_1, g_1$. The bottom rectangle in (9.21) commutes because of the naturality of $\mu_{X,FZ} : \phi_{X,FZ} \circ \psi_{X,FZ} \Rightarrow 1_{Mor_{\mathcal{X}}(X,GFZ)}$. Hence the outer rectangle of (9.21) commutes. The top rectangle of (9.22) commutes because of the naturality of $(\gamma^G)^{-1}, \mu_{X,FY}$, and $\mu_{Y,FZ}$ by comparing with the 2-cells of (9.18). The bottom rectangle of (9.22) commutes because it is the analogue of (9.19) for $f_2, g_2$. Hence the outer rectangle of (9.22) commutes. From (9.21) and (9.22) we conclude that both $F(\beta * \alpha) \odot \gamma^F_{f_1,g_1}$ and $\gamma^F_{f_2,g_2} \odot (F\beta * F\alpha)$ have $\phi_{X,FZ}$ images which fill in the right diagram of (9.23).

(9.23)

$$
\begin{array}{ccccc}
Fg_1 \circ Ff_1 & \qquad & G(Fg_1 \circ Ff_1) \circ \eta_X & \xRightarrow{(i_{\eta_Z} * \beta * \alpha) \odot \sigma_1} & \eta_Z \circ g_2 \circ f_2 \\
\Downarrow & & \Downarrow & & \| \\
F(g_2 \circ f_2) & & GF(g_2 \circ f_2) \circ \eta_X & \xRightarrow[\mu_{X,FZ}(\eta_Z \circ g_2 \circ f_2)]{} & \eta_Z \circ g_2 \circ f_2
\end{array}
$$

Since $\mu_{X,FZ}(\eta_Z \circ g_2 \circ f_2)$ is universal, we conclude that $F(\beta * \alpha) \odot \gamma^F_{f_1,g_1} = \gamma^F_{f_2,g_2} \odot (F\beta * F\alpha)$ and thus $\gamma^F$ is natural in its two variables.

We claim that $\delta^F$ and $\gamma^F$ satisfy the unit axiom for pseudo functors. Let $X \in Obj\ \mathcal{X}$ and let $f : X \to Y$ be a morphism in $\mathcal{X}$. We must show that $\gamma^F_{1_X,f} = (i_{Ff} * \delta^F_{X*})^{-1}$. By definition, $\gamma^F_{1_X,f}$ is the unique 2-cell $Ff \circ F1_X \Rightarrow F(f \circ 1_X)$ such that the composition of 2-cells

(9.24)

$$
\begin{array}{ccccc}
X & \xrightarrow{\eta_X} & GFX & \xrightarrow{G(Ff \circ F1_X)} & GFY \\
& \Downarrow i_{\eta_X} & & \Downarrow G(\gamma^F_{1_X,f}) & \\
X & \xrightarrow[\eta_X]{} & GFX & \xrightarrow[GF(f \circ 1_X)]{} & GFY \\
& & \Downarrow \mu_{X,FY}(\eta_Y \circ f \circ 1_X) & & \\
X & \xrightarrow[f \circ 1_X]{} & Y & \xrightarrow[\eta_Y]{} & GFY
\end{array}
$$

is the same as the composition of 2-cells

(9.25)

$$
\begin{array}{ccccccc}
X & \xrightarrow{\eta_X} & GFX & & \xrightarrow{G(Ff \circ F1_X)} & & GFY \\
& \Downarrow i_{\eta_X} & & & \Downarrow (\gamma^G_{F1_X,Ff})^{-1} & & \\
X & \xrightarrow[\eta_X]{} & GFX & \xrightarrow{GF1_X} & GFX & \xrightarrow{GFf} & GFY \\
& & \Downarrow \mu_{X,FX}(\eta_X \circ 1_X) & & & \Downarrow i_{GFf} & \\
X & \xrightarrow{1_X} & X & \xrightarrow[\eta_X]{} & GFX & \xrightarrow[GFf]{} & GFY \\
& \Downarrow i_{1_X} & & & \Downarrow \mu_{X,FY}(\eta_Y \circ f) & & \\
X & \xrightarrow[1_X]{} & X & \xrightarrow[f]{} & Y & \xrightarrow[\eta_Y]{} & GFY
\end{array}
$$

where universal 2-cells are drawn with dotted double arrows for clarity. We show that $(i_{Ff} * \delta^F_{X*})^{-1}$ is a 2-cell with this defining property for $\gamma^F_{1_X,f}$.

Since $\gamma^G$ is natural we can rewrite the first horizontal 2-cell composition in (9.25) as the composition of the first three 2-cells in the equal diagram (9.26).

(9.26)

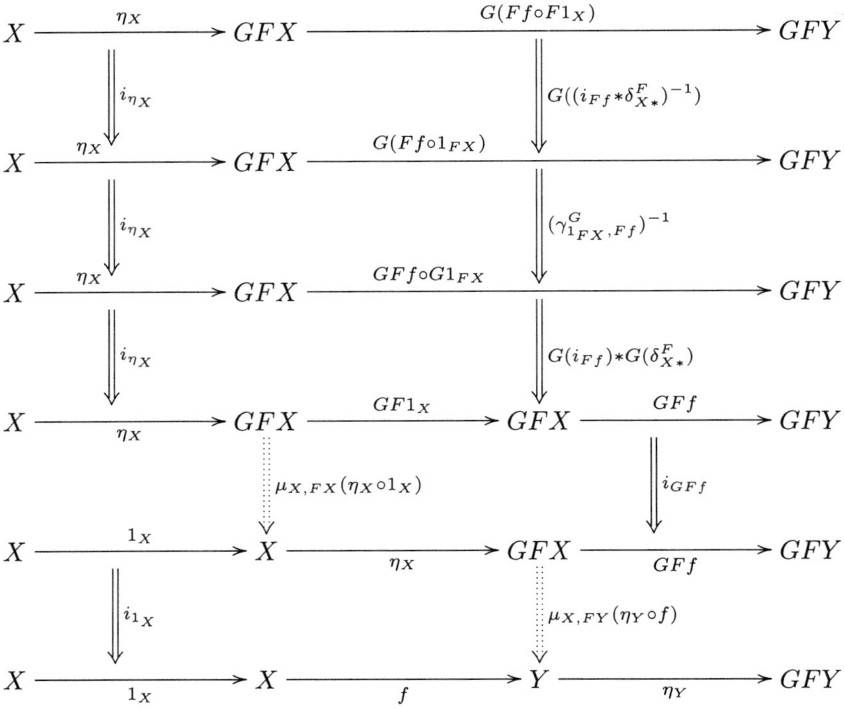

By the unit axiom for $G$, the definition of $\delta^F_{X*}$ in (9.17), and the interchange law we see that the second horizontal composition in (9.26) is

$$\begin{aligned}(\gamma^G_{1_{FX},Ff})^{-1} * i_{\eta_X} &= i_{GFf} * \delta^G_{FX*} * i_{\eta_X} \\ &= i_{GFf} * (\mu_{X,FX}(\eta_X \circ 1_X) \odot (G(\delta^F_{X*}) * i_{\eta_X}))^{-1} \\ &= (G(i_{Ff}) * G(\delta^F_{X*})^{-1} * i_{\eta_X}) \odot (i_{GFf} * \mu_{X,FX}(\eta_X \circ 1_X))^{-1}.\end{aligned}$$

Substituting this in (9.26) for $(\gamma^G_{1_{FX},Ff})^{-1} * i_{\eta_X}$ we see that the second horizontal composition in (9.26) cancels with the third and the fourth, leaving only

(9.27)

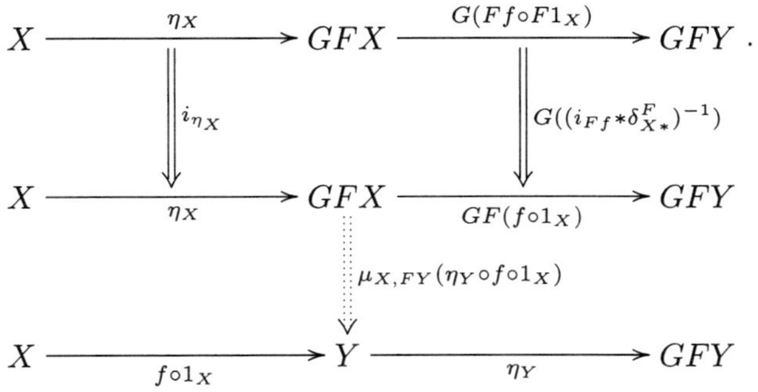

We see that the 2-cell compositions of (9.24),(9.25), (9.26), and (9.27) are all equal. Hence the 2-cell compositions (9.24) and (9.27) are equal and by universality of the 2-cell $\mu_{X,FY}(\eta_Y \circ f \circ 1_X)$ we have $\gamma^F_{1_X,f} = (i_{Ff} * \delta^F_{X*})^{-1}$. The other half of the unit axiom can be verified similarly. We conclude that $\delta^F$ and $\gamma^F$ satisfy the unit axiom for pseudo functors.

We claim that $\gamma^F$ satisfies the composition axiom for pseudo functors. Let $W \xrightarrow{f} X \xrightarrow{g} Y \xrightarrow{h} Z$ be morphisms of $\mathcal{X}$. We must show that $\gamma^F_{f,h\circ g} = \gamma^F_{g\circ f,h} \odot (i_{Fh} * \gamma^F_{f,g}) \odot (\gamma^F_{g,h} * i_{Ff})^{-1}$. By definition $(\gamma^F_{f,h\circ g})^{-1}$ is the unique 2-cell $F(h \circ g \circ f) \Rightarrow F(h \circ g) \circ Ff$ such that the composition of 2-cells

(9.28)

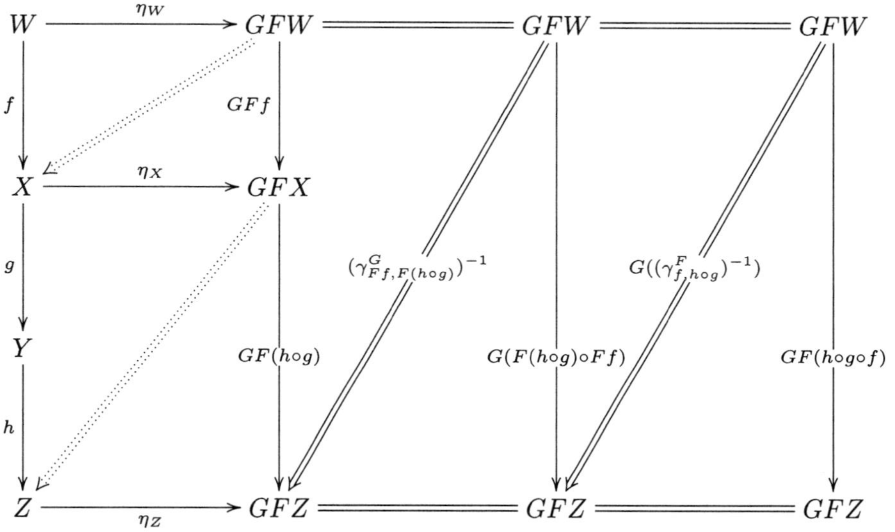

is the same as the universal 2-cell $\mu_{X,FZ}(\eta_Z \circ h \circ g \circ f)$. For clarity we continue to draw the universal 2-cells as dotted double arrows. We prove that replacing $(\gamma^F_{f,h\circ g})^{-1}$ in (9.28) by $(\gamma^F_{g\circ f,h} \odot (i_{Fh} * \gamma^F_{f,g}) \odot (\gamma^F_{g,h} * i_{Ff})^{-1})^{-1}$ still gives $\mu_{X,FZ}(\eta_Z \circ h \circ g \circ f)$. After that we conclude $\gamma^F_{f,h\circ g} = \gamma^F_{g\circ f,h} \odot (i_{Fh} * \gamma^F_{f,g}) \odot (\gamma^F_{g,h} * i_{Ff})^{-1}$ by the universality of the 2-cell $\mu_{X,FZ}(\eta_Z \circ h \circ g \circ f)$. To this end, we claim that the composition

(9.29)

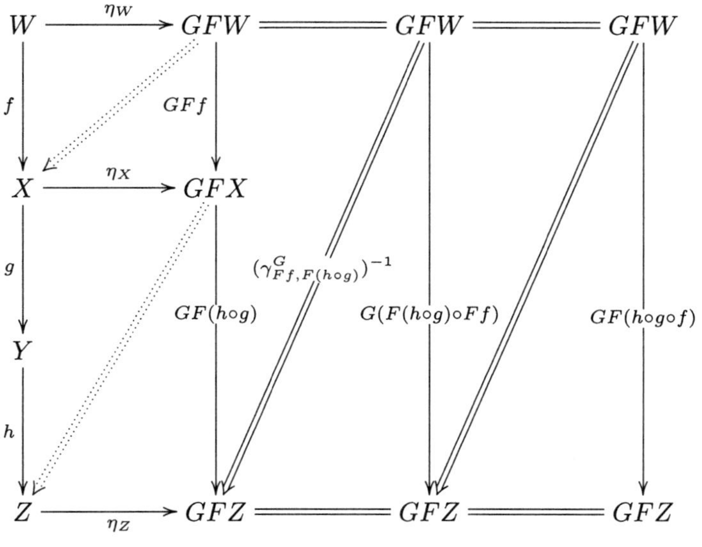

is the same as $\mu_{X,FZ}(\eta_Z \circ h \circ g \circ f)$, where the rightmost 2-cell is $G((\gamma^F_{g \circ f, h} \odot (i_{Fh} * \gamma^F_{f,g}) \odot (\gamma^F_{g,h} * i_{Ff})^{-1})^{-1})$. We do this by transforming (9.29) to a diagram known to be $\mu_{X,FZ}(\eta_Z \circ h \circ g \circ f)$. The naturality of $\gamma^G$ guarantees that

$$\begin{array}{ccc}
G(Fh \circ Fg) \circ GFf & \xrightarrow{\gamma^G_{Ff, Fh \circ Fg}} & G(Fh \circ Fg \circ Ff) \\
\Big\| G(\gamma^F_{g,h}) * i_{GFf} & & \Big\| G(\gamma^F_{g,h} * i_{Ff}) \\
GF(h \circ g) \circ GFf & \xrightarrow{\gamma^G_{Ff, F(h \circ g)}} & G(F(h \circ g) \circ Ff)
\end{array}$$

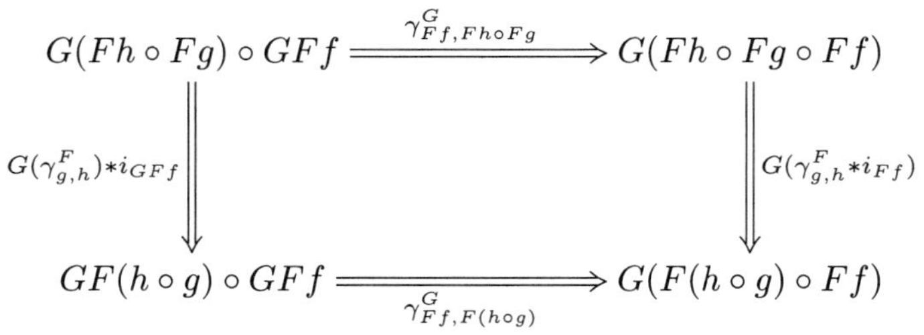

commutes. Using this commutivity to substitute for $(\gamma^G_{Ff, F(h \circ g)})^{-1}$ in (9.29) and cancelling $G(\gamma^F_{g,h} * i_{Ff})^{-1} \odot G(\gamma^F_{g,h} * i_{Ff})$ gives
(9.30)

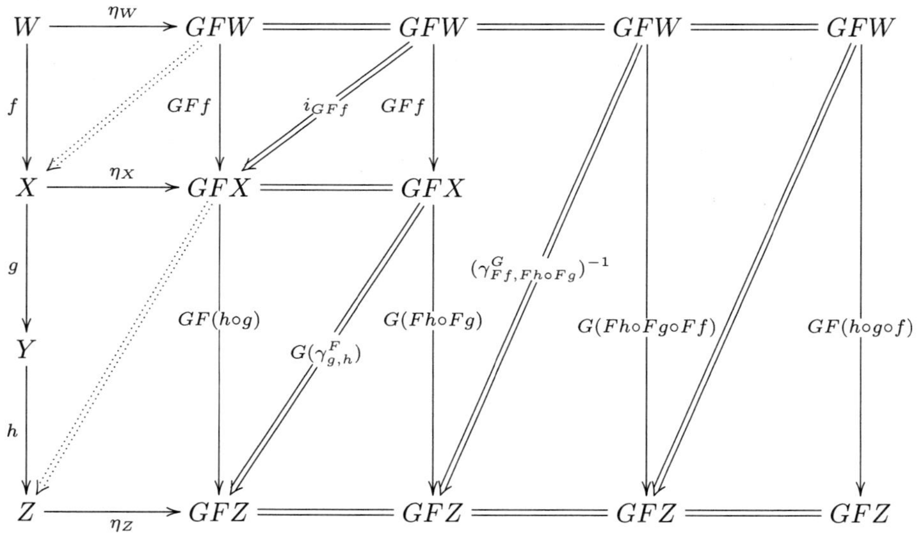

where the right 2-cell is $G((i_{Fh} * \gamma^F_{f,g})^{-1} \odot (\gamma^F_{g \circ f, h})^{-1})$. We have also implicitly used the fact that $G$ preserves the vertical composition of 2-cells. By the definition of $\gamma^F_{g,h}$ in (9.18) and (9.19), the lower left two rectangles of (9.30) can be rewritten to give the equal composition (9.31).

(9.31)

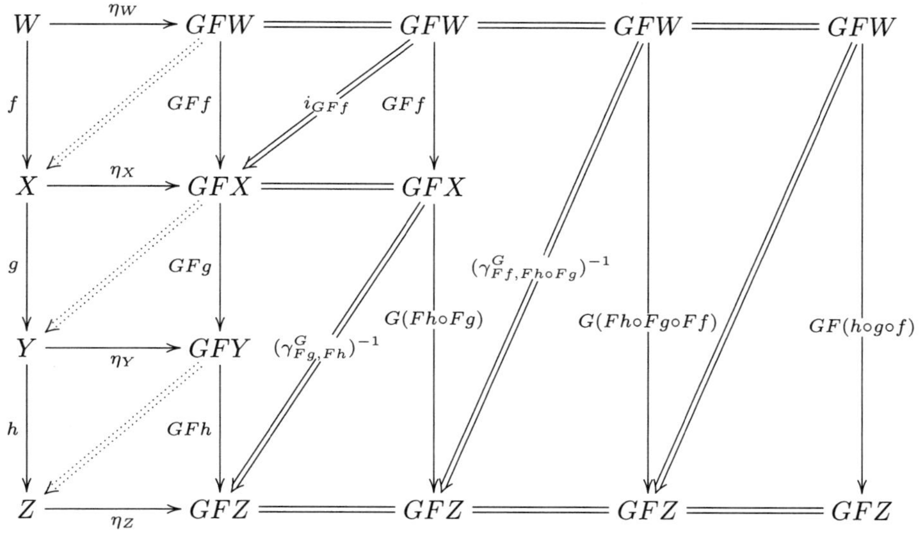


Recall that the composition axiom for the pseudo functor $G$ guarantees the commutivity of the following diagram.

$$\begin{array}{ccc} GFh \circ GFg \circ GFf & \xRightarrow{i_{GFh} * \gamma^G_{Ff,Fg}} & GFh \circ G(Fg \circ Ff) \\ {\scriptstyle \gamma^G_{Fg,Fh} * i_{GFf}} \Downarrow & & \Downarrow {\scriptstyle \gamma^G_{Fg\circ Ff,Fh}} \\ G(Fh \circ Fg) \circ GFf & \xRightarrow[\gamma^G_{Ff,Fh\circ Fg}]{} & G(Fh \circ Fg \circ Ff) \end{array}$$

Using this composition axiom for the pseudo functor $G$ we can replace the middle two columns of 2-cells in (9.31) to get the equal composition (9.32).

(9.32)

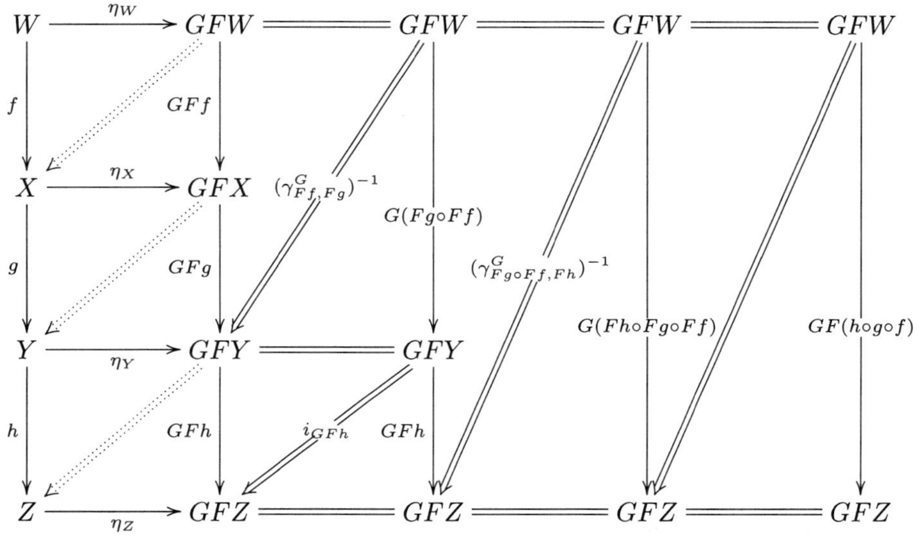

In (9.32) the right 2-cell is again $G((i_{Fh} * \gamma^F_{f,g})^{-1} \odot (\gamma^F_{g\circ f,h})^{-1})$ as in (9.30) and (9.31). By the definition of $\gamma^F_{f,g}$ in (9.18) and (9.19), we can rewrite the upper left three rectangles of (9.32) to obtain (9.33), which has $G((i_{Fh} * \gamma^F_{f,g})^{-1} \odot (\gamma^F_{g\circ f,h})^{-1})$ as its right 2-cell.

(9.33)

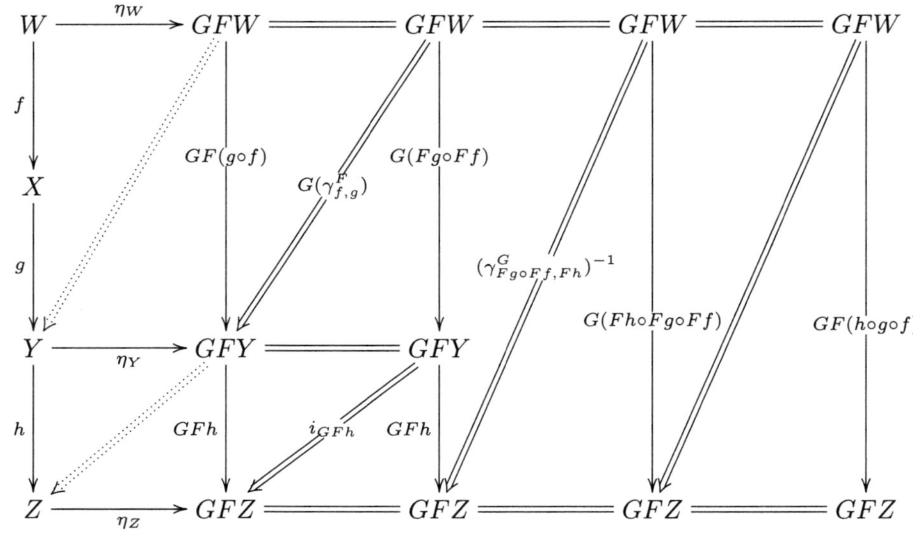

The naturality of $\gamma^G$ implies that the diagram

$$\begin{array}{ccc} GFh \circ G(Fg \circ Ff) & \xrightarrow{\gamma^G_{Fg\circ Ff, Fh}} & G(Fh \circ Fg \circ Ff) \\ {\scriptstyle G(i_{Fh})*G(\gamma^F_{f,g})} \Big\Downarrow & & \Big\Downarrow {\scriptstyle G(i_{Fh}*\gamma^F_{f,g})} \\ GFh \circ G(F(g\circ f)) & \xrightarrow[\gamma^G_{F(g\circ f), Fh}]{} & G(Fh \circ F(g\circ f)) \end{array}$$

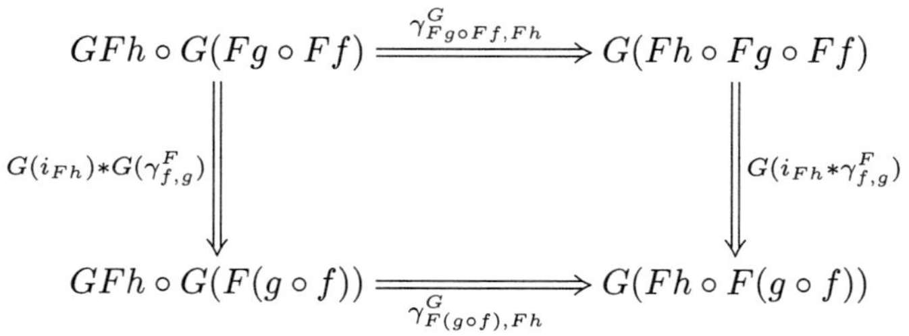

commutes. Using its commutivity, we can rewrite (9.33) by combining its middle two columns of 2-cells with $G((i_{Fh} * \gamma^F_{f,g})^{-1})$ from the last column to get (9.34).

(9.34)

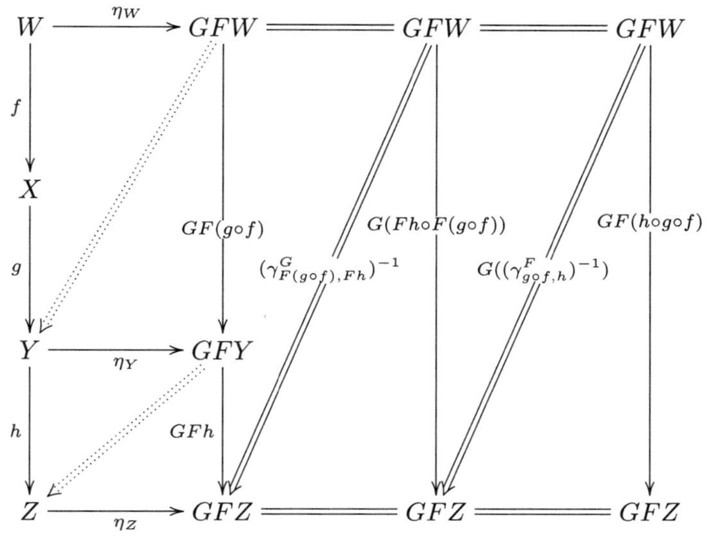


But by the definition of $\gamma^F_{g\circ f,h}$ in (9.18) and (9.19), the composition of 2-cells in (9.34) is precisely $\mu_{X,FZ}(\eta_Z \circ h \circ g \circ f)$. Since the compositions of 2-cells in the diagrams (9.29) through (9.34) are all equal, we conclude that the composition of 2-cells in (9.29) is $\mu_{X,FZ}(\eta_Z \circ h \circ g \circ f)$. We conclude that $\gamma^F_{f,h\circ g} = \gamma^F_{g\circ f,h} \odot (i_{Fh} * \gamma^F_{f,g}) \odot (\gamma^F_{g,h} * i_{Ff})^{-1}$ by the universality of $\mu_{X,FZ}(\eta_Z \circ h \circ g \circ f)$. Therefore $\gamma^F$ satisfies the composition axiom for pseudo functors.

In summary, we have constructed a pseudo functor $F : \mathcal{X} \to \mathcal{A}$ with natural coherence 2-cells $\delta^F$ and $\gamma^F$ and we have shown that they satisfy the unit axiom and composition axiom for pseudo functors.

Next we have to show that $F$ is a left biadjoint using Theorem 9.16. By hypothesis we already have a morphism $\eta_X : X \to G(FX)$ for all $X \in Obj\ \mathcal{X}$. We claim that the assignment $X \mapsto \eta_X$ is a pseudo natural transformation from $1_{\mathcal{X}}$ to $GF$. We need to define the 2-cells up to which $\eta$ is natural. For a morphism $f : X \to Y$ of $\mathcal{X}$ define $\tau_f := \mu_{X,FY}(\eta_Y \circ f)$. Then the diagram

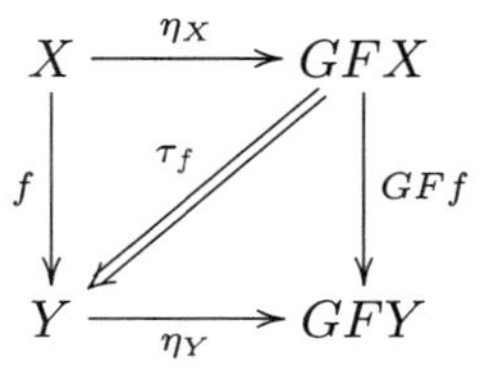


illustrates the source and target of the 2-cell. The map $f \mapsto \tau_f$ is natural because $\mu_{X,FY}$ is a natural transformation. More precisely let $\alpha : f_1 \Rightarrow f_2$ be a 2-cell in $\mathcal{X}$

and let $f_1, f_2 : X \to Y$ be morphisms in $\mathcal{X}$. Then

$$(9.35)\qquad \begin{array}{ccc} \phi_{X,FY}(\psi_{X,FY}(\eta_Y \circ f_1)) & \xRightarrow{\mu_{X,FY}(\eta_Y \circ f_1)} & \eta_Y \circ f_1 \\ \Big\Downarrow{\scriptstyle \phi_{X,FY}(\psi_{X,FY}(i_{\eta_Y} * \alpha))} & & \Big\Downarrow{\scriptstyle i_{\eta_Y} * \alpha} \\ \phi_{X,FY}(\psi_{X,FY}(\eta_Y \circ f_2)) & \xRightarrow[\mu_{X,FY}(\eta_Y \circ f_2)]{} & \eta_Y \circ f_2 \end{array}$$

commutes by the naturality of $\mu_{X,FY}$. By the definitions of $F$, $\tau_{f_1}$, and $\tau_{f_2}$, diagram (9.35) is the same as the diagram

$$(9.36)\qquad \begin{array}{ccc} GFf_1 \circ \eta_X & \xRightarrow{\tau_{f_1}} & \eta_Y \circ f_1 \\ \Big\Downarrow{\scriptstyle GF\alpha * i_{\eta_X}} & & \Big\Downarrow{\scriptstyle i_{\eta_Y} * \alpha} \\ GFf_2 \circ \eta_X & \xRightarrow[\tau_{f_2}]{} & \eta_Y \circ f_2 \end{array}$$

which says $f \mapsto \tau_f$ is natural. The map $f \mapsto \tau_f$ satisfies the unit axiom for pseudo natural transformations because of (9.17) and the definition of $\delta^{GF}$ for the composite pseudo functor $GF$. The map $f \mapsto \tau_f$ satisfies the composition axiom for pseudo natural transformations because of (9.18) and (9.19), and the definition of $\gamma^{GF}$ for the composite pseudo functor $GF$. Hence $\eta : 1_{\mathcal{X}} \Rightarrow GF$ is a pseudo natural transformation with coherence 2-cells $\tau$.

By Theorem 9.16, the constructed pseudo functor $F$ is a left biadjoint because $\eta : 1_{\mathcal{X}} \Rightarrow GF$ is a pseudo natural transformation such that $\eta_X : X \to G(FX)$ is a biuniversal arrow for all $X \in Obj\ \mathcal{X}$. □

We can summarize the previous two theorems in a way similar to Mac Lane's theorem on page 83 of [**39**] as follows.

THEOREM 9.18. *A biadjunction* $\langle F, G, \phi \rangle : \mathcal{X} \rightharpoonup \mathcal{A}$ *can be described up to pseudo natural pseudo isomorphism (defined below) by either of the following data:*

(1) *Pseudo functors*

$$\mathcal{X} \underset{G}{\overset{F}{\rightleftarrows}} \mathcal{A}$$

*and a pseudo natural transformation* $\eta : 1_{\mathcal{X}} \Rightarrow GF$ *such that each* $\eta_X : X \to G(FX)$ *is a biuniversal arrow from* $X$ *to* $G$. *Then* $\phi_{X,A}$ *is defined by* $\phi_{X,A}(f) = Gf \circ \eta_X$.

(2) *A pseudo functor* $G : \mathcal{A} \to \mathcal{X}$, *for each* $X \in Obj\ \mathcal{X}$ *an object* $R \in \mathcal{A}$ *depending on* $X$, *and for each* $X \in Obj\ \mathcal{X}$ *a biuniversal arrow* $\eta_X : X \to GR$ *from* $X$ *to* $G$. *Then the pseudo functor* $F$ *satisfies* $FX = R$ *on objects and there is a natural iso 2-cell* $GFh \circ \eta_X \Rightarrow \eta'_X \circ h$ *for morphisms* $h : X \to X'$.

*Proof:* Uniqueness will be proven below. □

Similar things can be formulated for bicounits. From 1-category theory we know that any two left adjoints to a functor are naturally isomorphic. A similar statement can be made for left biadjoints, although we need the concept of *pseudo natural pseudo isomorphism.*

DEFINITION 9.19. Let $F, F' : \mathcal{X} \to \mathcal{A}$ be pseudo functors. Then a pseudo natural transformation $\alpha : F \Rightarrow F'$ is called a *pseudo natural pseudo isomorphism* or *pseudo natural equivalence* if there exists a pseudo natural transformation $\alpha' : F' \Rightarrow F$ and there exist iso modifications $\alpha \odot \alpha' \rightsquigarrow 1_{F'}$ and $\alpha' \odot \alpha \rightsquigarrow 1_F$.

THEOREM 9.20. *Let $F, F' : \mathcal{X} \to \mathcal{A}$ be left biadjoints for a pseudo functor $G : \mathcal{A} \to \mathcal{X}$. Then there exists a pseudo natural pseudo isomorphism $\alpha : F \Rightarrow F'$*

*Proof:* For $X \in Obj\ \mathcal{X}$, let $\eta_X : X \to G(FX)$ and $\eta'_X : X \to G(F'X)$ be the biuniversal arrows obtained from the biadjunctions as in the theorems above. Then by Lemma 9.7 there exists a pseudo isomorphism $\alpha_X : FX \to F'X$ and a pseudo inverse $\alpha'_X : F'X \to FX$ as well as 2-cells $\alpha'_X \circ \alpha_X \Rightarrow 1_{FX}$ and $\alpha_X \circ \alpha'_X \Rightarrow 1_{F'X}$. It can be shown that the assignments $X \to \alpha_X$ and $X \to \alpha'_X$ are pseudo natural and the 2-cells determine modifications $\alpha' \odot \alpha \rightsquigarrow 1_F$ and $\alpha \odot \alpha' \rightsquigarrow 1_{F'}$.

For example, we construct the coherence 2-cell $\tau^\alpha$ up to which $\alpha$ is natural. For $f \in Mor_{\mathcal{X}}(X, Y)$ we have the following two diagrams.

(9.37)

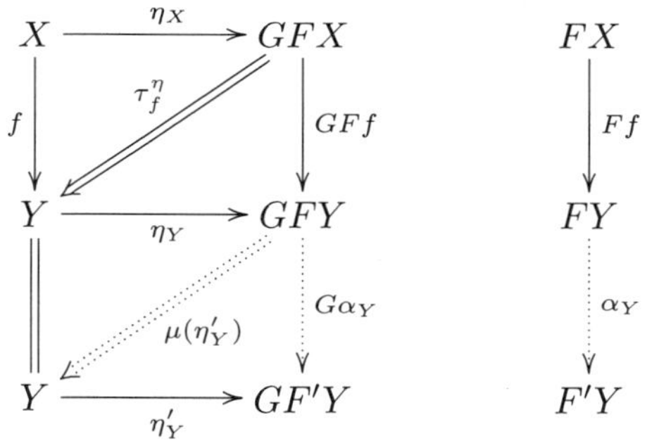


(9.38)

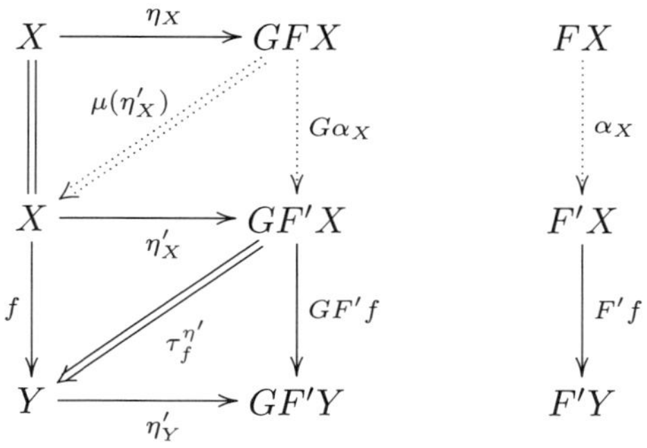


But they can also be filled in as

(9.39)

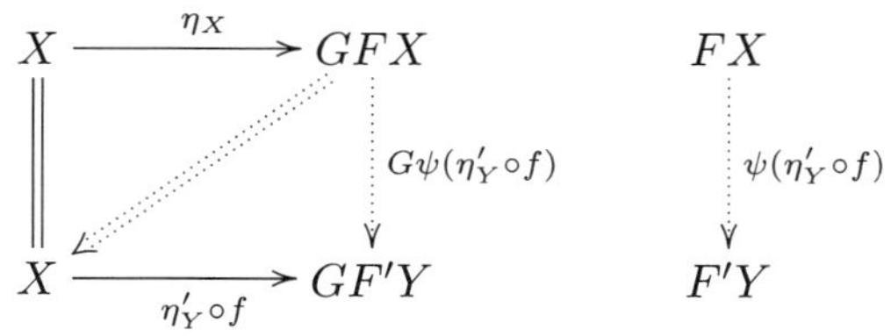

where the dashed 2-cell is universal. The universality gives us iso 2-cells $\nu_f$ and $\nu'_f$ as in

$$F'f \circ \alpha_X \xRightarrow{\nu'_f} \psi(\eta'_Y \circ f) \xLeftarrow{\nu_f} \alpha_Y \circ Ff$$

whose $\phi$ images factor (via the universal 2-cell) the 2-cells in (9.37) and (9.38) precomposed with the appropriate $(\gamma^G)^{-1}$'s. Define $\tau_f^\alpha := \tau_f := (\nu_f)^{-1} \odot \nu'_f$. This is the coherence 2-cell up to which $\alpha$ will be natural.

A sketch of the naturality of $f \mapsto \tau_f$ goes as follows. Let $\beta : f_1 \Rightarrow f_2$ be a 2-cell between $f_1, f_2 : X \to Y$. Then we must show that the outer rectangle of

$$\begin{array}{ccccc}
F'f_1 \circ \alpha_X & \xRightarrow{\nu'_{f_1}} & \psi(\eta'_Y \circ f_1) & \xLeftarrow{\nu_{f_1}} & \alpha_Y \circ Ff_1 \\
{\scriptstyle F'\beta * i_{\alpha_X}} \Downarrow & & \Downarrow {\scriptstyle \psi(i_{\eta'_Y} * \beta)} & & \Downarrow {\scriptstyle i_{\alpha_Y} * F\beta} \\
F'f_2 \circ \alpha_X & \xRightarrow[\nu'_{f_2}]{} & \psi(\eta'_Y \circ f_2) & \xLeftarrow[\nu_{f_2}]{} & \alpha_Y \circ Ff_2
\end{array}$$

commutes. We do this by showing that the individual inner squares commute by applying $\phi$ and using the universality and the fact that $\mu$ is a natural isomorphism. It also involves the naturality of the $\gamma^G$'s.

We can also show that $\tau$ satisfies the composition and unit axiom, although it is lengthy. Lastly we must verify that the 2-cell assignments at the start actually give modifications $\alpha' \odot \alpha \rightsquigarrow 1_F$ and $\alpha \odot \alpha' \rightsquigarrow 1_{F'}$.

Thus, any two left biadjoints are pseudo naturally pseudo isomorphic. □

There is a relationship between bi(co)limits and biadjoints, just like for (co)limits and adjoints.

REMARK 9.21. Let $\mathcal{C}$ be a 2-category which admits bicolimits and bilimits and let $\mathcal{J}$ be a 1-category. Let $\mathcal{C}^{\mathcal{J}}$ be the 2-category with objects pseudo functors $\mathcal{J} \to \mathcal{C}$, morphisms pseudo natural transformations, and 2-cells the modifications. Let $\Delta : \mathcal{C} \to \mathcal{C}^{\mathcal{J}}$ be the diagonal 2-functor. Then *bicolim* $: \mathcal{C}^{\mathcal{J}} \to \mathcal{C}$ is a left biadjoint for $\Delta$ and the arrows of the biunit constructed in Theorem 9.16 are the universal pseudo cones. Similarly, *bilim* $: \mathcal{C}^{\mathcal{J}} \to \mathcal{C}$ is a right biadjoint for $\Delta$ and the arrows of the bicounit are the universal pseudo cones.

CHAPTER 10

# Forgetful 2-Functors for Pseudo Algebras

Next we show that forgetful 2-functors for pseudo algebras admit left biadjoints. Let us consider the strict case as an example of what we do below. Let $S$ be the theory of abelian groups and let $T$ be the theory of rings. Then we have an inclusion $S \hookrightarrow T$. Let $X$ be a discrete $T$-algebra, *i.e.* $X$ is a set and we have a morphism of theories $T \to End(X)$. Then $X$ can be made into an $S$-algebra by the composite map of theories $S \hookrightarrow T \to End(X)$. This precomposition with the inclusion arrow forgets the ring structure on the set $X$ and results in the underlying abelian group. This precomposition with the inclusion defines the *forgetful functor* from the category of rings to the category of abelian groups. It admits a left adjoint which is the appropriate *free functor*. Similarly, for any morphism of theories $S \to T$ we have a forgetful 2-functor from pseudo $T$-algebras to pseudo $S$-algebras and this 2-functor admits a left biadjoint. Blackwell, Kelly, and Power have shown that left biadjoints exist for the analogous 2-functors on 2-categories of strict algebras over 2-monads with pseudo morphisms in [**9**]. Lack has given sufficient conditions in [**33**] under which the inclusion of strict algebras over a 2-monad into pseudo algebras over the same 2-monad admits a left adjoint whose unit has components that are equivalences. In such cases, every pseudo algebra over the 2-monad is equivalent to a strict algebra over the 2-monad. Yanofsky has also studied quasiadjoints to forgetful 2-functors induced by morphisms of 2-theories in [**54**], although his 2-theories are different from those of [**25**], [**26**], [**27**], and Chapter 13.

DEFINITION 10.1. Let $\phi : S \to T$ be a morphism of theories and let $X$ be a pseudo $T$-algebra with structure maps $\Psi_n : T(n) \to End(X)(n)$. Let $UX$ be the pseudo $S$-algebra which has $X$ as its underlying category and $S$ structure maps defined by $\Psi_n(\phi(w)) : X^n \to X$ for $w \in S(n)$. Defining $U$ analogously for morphisms and 2-cells of the 2-category of pseudo $T$-algebras yields a strict 2-functor $U$ from the 2-category of pseudo $T$-algebras to the 2-category of pseudo $S$-algebras called the *forgetful 2-functor associated to $\phi$.*

To show that the forgetful 2-functor associated to $\phi$ admits a left biadjoint, we need to find a biuniversal arrow of the following type: given a pseudo $S$-algebra $X$ there should exist a pseudo $T$-algebra $R$ and a biuniversal arrow $\eta_X : X \to UR$ in the category of pseudo $S$-algebras. We define this $R$ now.

NOTATION 10.2. Let $T$ be a theory. Let $T'$ denote the free theory on the sequence of sets $T(0), T(1), \ldots$ underlying the theory $T$. The category $Alg'$ is the category whose objects are small $T'$-algebras and whose morphisms are morphisms of strict $T'$-algebras. Let $Obj\ Graph'$ be the collection of small directed graphs whose object sets are discrete $T'$-algebras. Let $Mor\ Graph'$ be the collection of morphisms of directed graphs whose object components are morphisms of discrete

$T'$-algebras. Then *Graph'* is a category. We denote by $V'$ the left adjoint to the forgetful functor $V : Alg' \to Graph'$.

The forgetful functor $V : Alg' \to Graph'$ admits a left adjoint $V' : Graph' \to Alg'$ by Freyd's Adjoint Functor Theorem. The functor $V'$ is similar to taking the free category on a directed graph, except the resulting category is also a $T'$-algebra. The objects of the underlying directed graph of $V'Y$ and the objects of the directed graph $Y$ are the same.

DEFINITION 10.3. Let $\phi : S \to T$ be a morphism of theories. Let $X$ be a pseudo $S$-algebra with structure maps $\Psi_n : S(n) \to End(X)(n)$. We define the *free pseudo $T$-algebra $R$ on the pseudo $S$-algebra $X$ associated to $\phi$* via intermediate steps $R_{G'}$ and $R'$ as follows. Let $Obj\ R_{G'}$ be the (discrete) free $T'$-algebra on the discrete category $Obj\ X$ and let $Mor\ R_{G'}$ be the collection of the following arrows:

(1) For every $n \in \mathbf{N}$, for all words $w \in T(n)$, $w_1 \in T(m_1), \ldots, w_n \in T(m_n)$, and for all objects $A_1^1, \ldots, A_{m_1}^1, A_1^2, \ldots, A_{m_2}^2, \ldots, A_1^n, \ldots, A_{m_n}^n \in Obj\ R_{G'}$ there are arrows

$$c_{w,w_1,\ldots,w_n}(A_1^1, \ldots, A_{m_n}^n) :$$

$$w \circ (w_1, \ldots, w_n)(A_1^1, \ldots, A_{m_n}^n) \longrightarrow w(w_1(A_1^1, \ldots, A_{m_1}^1), \ldots, w_n(A_1^n, \ldots, A_{m_n}^n))$$

$$c_{w,w_1,\ldots,w_n}^{-1}(A_1^1, \ldots, A_{m_n}^n) :$$

$$w(w_1(A_1^1, \ldots, A_{m_1}^1), \ldots, w_n(A_1^n, \ldots, A_{m_n}^n)) \longrightarrow w \circ (w_1, \ldots, w_n)(A_1^1, \ldots, A_{m_n}^n).$$

Here $w \circ (w_1, \ldots, w_n)$ is the composition in the original theory $T$. The target $w(w_1(A_1^1, \ldots, A_{m_1}^1), \ldots, w_n(A_1^n, \ldots, A_{m_n}^n))$ is the result of composing in the free theory and applying it to the $A$'s in the free algebra.

(2) For every $A \in Obj\ R_{G'}$ there are arrows

$$I_A : 1(A) \longrightarrow A$$

$$I_A^{-1} : A \longrightarrow 1(A).$$

Here 1 is the unit of the original theory $T$.

(3) For every word $w \in T(m)$, for every function $f : \{1, \ldots, m\} \to \{1, \ldots, n\}$, and for all objects $A_1, \ldots, A_n \in Obj\ R_{G'}$ there are arrows

$$s_{w,f}(A_1, \ldots, A_n) : w_f(A_1, \ldots, A_n) \longrightarrow w(A_{f1}, \ldots, A_{fm})$$

$$s_{w,f}^{-1}(A_1, \ldots, A_n) : w(A_{f1}, \ldots, A_{fm}) \longrightarrow w_f(A_1, \ldots, A_n).$$

The substituted word $w_f$ is the substituted word in the original theory $T$. The target $w(A_{f1}, \ldots, A_{fm})$ is the result of substituting in $w$ in the free theory and then evaluating on the $A$'s.

(4) For every word $w \in S(n)$ and objects $A_1, \ldots, A_n$ of $X$ there are arrows

$$\rho_w^{\eta}(A_1, \ldots A_n) : \Psi(w)(A_1, \ldots, A_n) \longrightarrow \phi(w)(A_1, \ldots, A_n)$$

$$\rho_w^{\eta-1}(A_1, \ldots A_n) : \phi(w)(A_1, \ldots, A_n) \longrightarrow \Psi(w)(A_1, \ldots, A_n).$$

(5) Include also all elements of $Mor\ X$.

Then $R_{G'}$ is an object of *Graph'*. Now we apply $V'$ to $R_{G'}$ and we get a category $R'$ which is a $T'$-algebra. The objects of $R_{G'}$ and $R'$ are the same.

Let $K$ be the smallest congruence on $R'$ with the following properties:

(1) All of the relations necessary to make the coherence arrows (including $\rho^\eta_w$) into natural transformations belong to $K$. For example, if $A, B \in Obj\ R'$ and $f : A \to B$ is a morphism of $R'$, then the relation $I_A \circ f = 1(f) \circ I_B$ belongs to $K$.
(2) All of the relations necessary to make the coherence arrows (including $\rho^\eta_w$) into isos are in $K$. For example, for every $A \in Obj\ R'$ the relations $I_A \circ I_A^{-1} = 1_A$ and $I_A^{-1} \circ I_A = 1_A$ are in $K$.
(3) All of the relations for pseudo algebras listed in Definition 7.1 belong to $K$, where the objects range over the objects of $R'$.
(4) The original composition relations in the category $X$ belong to $K$.
(5) The coherence diagrams necessary to make the inclusion $\eta_X : X \to UR$ into a morphism of pseudo $S$-algebras are in $K$. These diagrams are listed in Definition 7.4. Note that these coherence diagrams will involve the arrows $\rho^\eta_w(A_1, \ldots, A_n) : \Psi(w)(A_1, \ldots, A_n) \to \phi(w)(A_1, \ldots, A_n)$ for $w \in S(n)$ and objects $A_1, \ldots, A_n \in Obj\ X$.
(6) If the relations $f_1 = g_1, \ldots, f_n = g_n$ are in $K$ and $w \in T'(n)$, then the relation $w(f_1, \ldots, f_n) = w(g_1, \ldots g_n)$ is also in $K$.

Next mod out by the congruence $K$ in $R'$ to obtain the quotient category $R$ called the *free pseudo $T$-algebra on the pseudo $S$-algebra $X$ associated to $\phi$*. We do not use a capital Greek letter to denote the structure maps of the pseudo $T$-algebra $R$. Instead we write the words directly.

In all of the following lemmas in this chapter we use the notation just introduced in Definition 10.1, Notation 10.2, and Definition 10.3.

LEMMA 10.4. *In the notation of the previous definition, the free pseudo $T$-algebra $R$ on the pseudo $S$-algebra $X$ associated to $\phi$ is a pseudo $T$-algebra.*

*Proof:* First we note that $R$ is a (strict) $T'$-algebra. The functor from the word $w \in T'(n)$ induces a functor on the quotient by relation 6 and the composition and identities in $T'$ are preserved. The structure maps have the coherence isos required of a pseudo $T$-algebra because of the arrows we threw in. The coherence isos satisfy the required coherence diagrams because of relations 1 and 2. Hence $R$ is a pseudo $T$-algebra. □

LEMMA 10.5. *The inclusion functor denoted $\eta_X : X \to UR$ is a morphism of pseudo $S$-algebras.*

*Proof:* The inclusion is a functor because of relation 4. It is a morphism of pseudo $S$-algebras because for all $w \in S(n)$ the natural transformation $\rho^\eta_w : \eta_X \circ \Psi(w) \Rightarrow \phi(w)(\eta_X, \ldots, \eta_X)$ satisfies the required coherences by the relations in 1. and 5. □

LEMMA 10.6. *For every pseudo $T$-algebra $D$ and every morphism $H : X \to UD$ of pseudo $S$-algebras, there exists a morphism $H' : R \to D$ of pseudo $T$-algebras* *such that*

$$\begin{array}{ccccc} X & \xrightarrow{\eta_X} & UR & & R \\ \| & & \vdots\, UH' & & \vdots\, H' \\ X & \xrightarrow[H]{} & UD & & D \end{array}$$

*commutes.*

*Proof:* Let $\Phi$ denote the structure maps of the pseudo $T$-algebra $D$. As above, $\Psi$ denotes the structure maps of the pseudo $S$-algebra $X$ and we suppress the capital Greek letter when denoting the structure maps of the pseudo $T$-algebra $R$. Note that $D$ is a strict $T'$-algebra and we can therefore apply the forgetful 2-functor $V : Alg' \to Graph'$ to it. We also use $\Phi$ to denote the structure maps of the strict $T'$ algebra $D$. To construct the morphism $H'$, we define a morphism $H_0' : R_{G'} \to VD$ in $Graph'$, which induces a morphism $H_1' : R' \to D$ in $Alg'$ by the definition of the left adjoint to $V$. Then we show that $H_1'$ preserves the congruence $K$ and therefore induces a functor $H' : R \to D$. Lastly we show that $H'$ is a morphism of pseudo $T$-algebras such that the desired diagram commutes.

We now define a morphism $H_0' : R_{G'} \to VD$ in $Graph'$. Defining $H_0'A := HA$ for $A \in Obj\ X$ induces a map $H_0' : Obj\ R_{G'} \to Obj\ D$ of discrete $T'$ algebras. For $f \in Mor\ X$ define $H_0'f := Hf$. For every $w \in S(n)$ and objects $A_1, \ldots, A_n \in Obj\ X$ let $H_0'$ map the arrows $\rho_w^\eta(A_1, \ldots A_n) : \Psi(w)(A_1, \ldots, A_n) \to \phi(w)(A_1, \ldots, A_n)$ to the coherence isos $\rho_w^H(A_1, \ldots A_n) : H(\Psi(w)(A_1, \ldots, A_n)) \to \Phi(\phi(w))(HA_1, \ldots, HA_n)$. Note that the source and target of $\rho_w^H(A_1, \ldots A_n)$ are equal to $H_0'(\Psi(w)(A_1, \ldots, A_n))$ and $H_0'(\phi(w)(A_1, \ldots, A_n))$ respectively. Let $H_0'$ map the other coherence arrows 1 through 3. to the analogous ones in $Mor\ D$ with $H_0$ applied to sources and targets. Thus we have defined a morphism $H_0' : R_{G'} \to VD$ in $Graph'$.

The morphism $H_0' : R_{G'} \to VD$ in $Graph'$ induces a morphism $H_1' : R' \to D$ of $Alg'$ by the definition of the left adjoint to $V$. We claim that $H_1'$ preserves the congruence $K$. It suffices to check the relations 1 through 6. We verify them in order of the list above.

(1) These are satisfied because the analogous arrows for $D$ and $H$ are natural transformations and $H_1'$ maps coherence arrows to coherence arrows.
(2) These are satisfied because the analogous arrows for $D$ and $H$ are isos and $H_1'$ maps coherence arrows to coherence arrows.
(3) The target category $D$ is a pseudo $T$-algebra so these are satisfied.
(4) The functor $H$ preserves the relations of the category $X$ and $H_1'$ is defined in terms of $H$, which implies that these are satisfied.
(5) These are satisfied because $\rho_w^H$ satisfies the coherences and $H_1'(\rho_w^\eta) = \rho_w^H$.
(6) This is by induction. The base case is showing 1 through 5. as was just done. Suppose the relations $f_1 = g_1, \ldots, f_n = g_n$ are in $K$ and

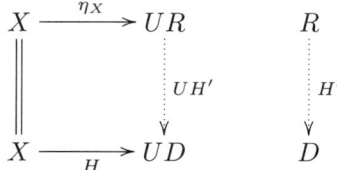

$H_1'f_i = H_1'g_i$ for all $i = 1, \ldots, n$. That is our induction hypothesis. Then

$$\begin{aligned} H_1'(w(f_1, \ldots, f_n)) &= \Phi(w)(H_1'(f_1), \ldots, H_1'(f_n)) \text{ since } H_1' \\ &\qquad \text{is a morphism of } T'\text{-algebras} \\ &= \Phi(w)(H_1'g_1, \ldots, H_1'g_n) \text{ by induction hypothesis} \\ &= H_1'(w(g_1, \ldots, g_n)) \text{ since } H_1' \\ &\qquad \text{is a morphism of } T'\text{-algebras.} \end{aligned}$$

Thus $H_1'(w(f_1, \ldots, f_n)) = H_1'(w(g_1, \ldots, g_n))$ and $H_1'$ satisfies this relation.

Since $H_1'$ satisfies the relations, we conclude that $H_1' : R' \to D$ induces a functor $H' : R \to D$ such that $H_1' = H' \circ Q$ where $Q : R' \to R$ is the projection functor onto the quotient category. The functor $H' : R \to D$ is a morphism of strict $T'$-algebras because for $w \in T'(n)$, $A_1, \ldots, A_n \in Obj\ R$, and for morphisms $f_1, \ldots, f_n \in Mor\ R$ we have

$$\begin{aligned} H'(w(A_1, \ldots, A_n)) &= H_1'(w(A_1, \ldots, A_n)) \\ &= \Phi(w)(H_1'A_1, \ldots, H_1'A_n) \\ &= \Phi(w)(H'A_1, \ldots, H'A_n) \end{aligned}$$

since $H_1'$ and $H'$ agree on objects. We also have

$$\begin{aligned} H'(w(f_1, \ldots, f_n)) &= H_1'(w(f_1, \ldots, f_n)) \\ &= \Phi(w)(H_1'f_1, \ldots, H_1'f_n) \\ &= \Phi(w)(H'f_1, \ldots, H'f_n) \end{aligned}$$

where $H_1'$ is actually applied to representatives of $w(f_1, \ldots, f_n), f_1, \ldots, f_n$. Hence $H'$ is a morphism of strict $T'$-algebras and also a morphism of pseudo $T$-algebras, since $T(n) \subseteq T'(n)$ although this inclusion is not necessarily a map of theories. According to these two demonstrations, the coherence 2-cells for the morphism $H'$ of pseudo $T$-algebras are just identities.

We claim that

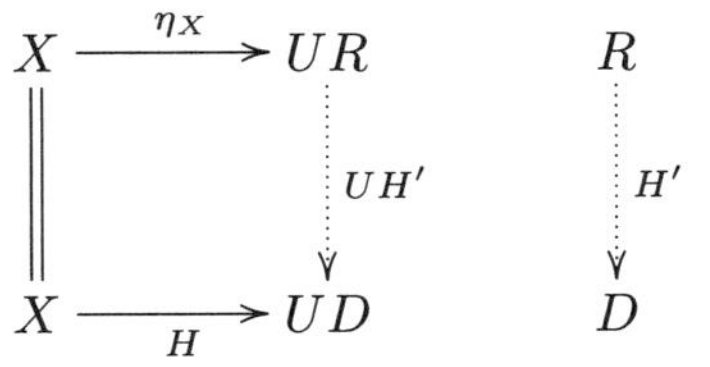


commutes. It is sufficient to check this for the underlying functors and the coherence 2-cells. The underlying functor of $H'$ is the same as the underlying functor of $UH'$. Let $A \in Obj\ X$. Then $UH' \circ \eta_X(A) = UH'(A) = H'A = HA$. Similarly, for $f \in Mor\ X$ we have $UH' \circ \eta_X(f) = UH'(f) = H'f = Hf$. Hence the diagram commutes. The coherence 2-cells also commute because $H'(\rho_w^\eta) = \rho_w^H$ and because the coherence 2-cells of $H'$ are identities. $\square$

LEMMA 10.7. *The inclusion morphism $\eta_X : X \to UR$ is a biuniversal arrow from $X$ to the forgetful 2-functor.*

*Proof:* Let $D$ be a pseudo $T$-algebra. Let $Mor_S(X, UD)$ denote the category of morphisms of pseudo $S$-algebras from $X$ to $UD$. Let $Mor_T(R, D)$ denote the category of morphisms of pseudo $T$-algebras from $R$ to $D$. Let $\phi : Mor_T(R, D) \to Mor_S(X, UD)$ be the functor defined by $H' \mapsto UH' \circ \eta_X$ and $\gamma \mapsto U\gamma * i_{\eta_X}$. Define a functor $\psi : Mor_S(X, UD) \to Mor_T(R, D)$ as follows. For $H \in Obj\ Mor_S(X, UD)$ let $\psi H := H'$ where $H' : R \to D$ is the morphism of pseudo $T$ algebras constructed in the previous lemma.

If $H, J \in Obj\ Mor_S(X, UD)$ and $\beta : H \Rightarrow J$ is a 2-cell in the 2-category of pseudo $S$-algebras, define $\psi(\beta) = \beta' : H' \Rightarrow J'$ inductively as follows. If $A \in Obj\ X$ then define $\beta' A$ to make

$$\begin{array}{ccc} H'A & \xrightarrow{\beta' A} & J'A \\ \| & & \| \\ HA & \xrightarrow[\beta A]{} & JA \end{array}$$

commute. If $w \in T'(n)$ and $\beta'$ is already defined for $A_1, \ldots, A_n \in Obj\ R$, then $\beta'(w(A_1, \ldots, A_n)) := \Phi(w)(\beta' A_1, \ldots, \beta' A_n)$. The following inductive proof shows that $\beta' : H' \Rightarrow J'$ is a natural transformation. For $f \in Mor\ X$ the naturality of $\beta'$ is guaranteed by the naturality of $\beta : H \Rightarrow J$. The naturality of $\beta'$ for the coherence isos thrown into the category $R$ during its construction follows because $H'$ and $J'$ take coherence isos of $R$ to analogous ones in $D$ and the coherences isos in $D$ are natural. That concludes the base case for the induction. Now suppose $\beta'$ is natural for morphisms $f_i \in Mor_R(A_i, B_i)$ for $i = 1, \ldots, n$ and $w \in T'(n)$. Then

$$\begin{array}{ccc} H'w(A_1, \ldots, A_n) & \xrightarrow{\beta' w(A_1, \ldots, A_n)} & J'w(A_1, \ldots, A_n) \\ {\scriptstyle H'w(f_1, \ldots, f_n)} \downarrow & & \downarrow {\scriptstyle J'w(f_1, \ldots, f_n)} \\ H'w(B_1, \ldots, B_n) & \xrightarrow[\beta' w(B_1, \ldots, B_n)]{} & J'w(B_1, \ldots, B_n) \end{array}$$

commutes because $w$ commutes with everything in the diagram by definition and because we apply the functor $\Phi(w)$ to each of the individual naturality diagrams for $f_i : A_i \to B_i$ and $i = 1, \ldots, n$. Hence $\beta'$ is natural for any morphism in $R$ by this inductive proof. Moreover, the natural transformation commutes appropriately with $\rho^{H'}$ and $\rho^{J'}$ because they are trivial and $\beta'(w(A_1, \ldots, A_n) = \Phi(w)(\beta' A_1, \ldots, \beta' A_n)$. Hence $\psi(\beta) = \beta'$ is a 2-cell in the 2-category of pseudo $T$-algebras.

It is routine to check inductively that the assignment $\psi : Mor_S(X, UD) \to Mor_T(R, D)$ preserves identities and compositions and is thus a functor.

We claim that $\psi$ is a right adjoint for $\phi$. By the previous lemma $\phi \circ \psi(H) = H$ for all $H \in Obj\ Mor_S(X, UD)$. We easily see that $\phi \circ \psi(\beta) = \beta$ for all $\beta \in Mor\ Mor_S(X, UD)$. Hence the counit $\mu : \phi \circ \psi \Rightarrow 1_{Mor_S(X,UD)}$ is the identity natural transformation, which is of course a natural isomorphism. Next we define a unit $\theta : 1_{Mor_T(R,D)} \Rightarrow \psi \circ \phi$. For $J' \in Mor_T(R, D)$ let $H' := \psi \circ \phi(J')$. Recall that $H'$ is strict, *i.e.* $\rho^{H'}$ is trivial, while $J'$ may not be strict. We define a 2-cell $\theta(J') : J' \Rightarrow H' = \psi \circ \phi(J')$ in the category of pseudo $T$-algebras inductively. For $A \in Obj\ X \subseteq Obj\ R$ set $\theta(J')(A) := 1_{J'A}$. Suppose $w \in T'(n)$ and $\theta(J')$ is already defined for $A_1, \ldots, A_n \in Obj\ R$. Then define $\theta(J')(w(A_1, \ldots, A_n) : J'(w(A_1, \ldots, A_n)) \to H'(w(A_1, \ldots, A_n))$ by $\Phi(w)(\theta(J')A_1, \ldots, \theta(J')A_n) \circ \rho_w^{J'}(A_1, \ldots, A_n)$. An inductive proof, similar to the one above but also using the naturality of $\rho_w^{J'}$, shows that

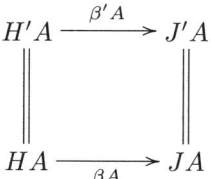

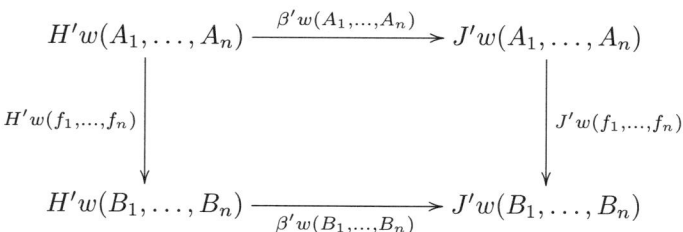

$\theta(J')$ is a natural transformation and commutes with $\rho^{J'}$ and $\rho^{H'}$ appropriately, *i.e.* $\theta(J') : J' \Rightarrow H'$ is a 2-cell. It is also iso by induction. The assignment $J' \mapsto \theta(J')$ is natural by an inductive argument that uses the diagram in the definition of 2-cell in the 2-category of pseudo $T$-algebras. Hence $\theta : 1_{Mor_T(R,D)} \Rightarrow \psi \circ \phi$ is a natural isomorphism. If we can show that $\theta$ and $\mu$ satisfy the triangular identities, then we can conclude that $\psi$ is a right adjoint for $\phi$

We claim that the unit $\theta$ and the counit $\mu$ satisfy the triangular identities. First we show that

$$\psi \xRightarrow{\theta * i_\psi} \psi \circ \phi \circ \psi \xRightarrow{i_\psi * \mu} \psi \tag{10.1}$$

is the identity natural transformation $i_\psi : \psi \Rightarrow \psi$. Let $H \in Obj\ Mor_S(X, UD)$. Then

$$\begin{aligned}(i_\psi * \mu) \odot (\theta * i_\psi)(H) &= \psi(\mu_H) \circ \theta_{\psi H} \text{ by definition} \\ &= \theta_{\psi H} \text{ since } \mu_H \text{ is trivial.}\end{aligned}$$

But $\theta_{\psi H} = \theta(\psi H)$ is the trivial 2-cell $\psi H \Rightarrow \psi H$ because $\psi H$ is a strict morphism of pseudo $T$-algebras, *i.e.* $\rho^{\psi H}_w$ is trivial. Hence (10.1) is $i_\psi : \psi \Rightarrow \psi$. Next we show that

$$\phi \xRightarrow{i_\phi * \theta} \phi \circ \psi \circ \phi \xRightarrow{\mu * i_\phi} \phi \tag{10.2}$$

is the identity natural transformation $i_\phi : \phi \Rightarrow \phi$. Let $J' \in Obj\ Mor_T(R, D)$. Then

$$\begin{aligned}(\mu * i_\phi) \odot (i_\phi * \theta)(J') &= \mu_{\phi J'} \circ \phi(\theta_{J'}) \text{ by definition} \\ &= \phi(\theta_{J'}) \text{ since } \mu_{\phi J'} \text{ is trivial} \\ &= \theta_{J'} * i_{\eta_X} \text{ by definition.}\end{aligned}$$

But $\theta_{J'} * i_{\eta_X}$ is the trivial 2-cell $\phi(J') = J' \circ \eta_X \Rightarrow J' \circ \eta_X$ because $\theta_{J'}(A) = \theta(J')(A) = 1_{J'A}$ for all $A \in Obj\ X$ and $\eta_X : X \to R$ is the inclusion functor. Hence (10.2) is the identity natural transformation $i_\phi : \phi \Rightarrow \phi$. Thus the unit and counit satisfy the triangular identities and $\psi$ is a right adjoint for $\phi$. Moreover, $\phi$ is an equivalence because the unit and counit are natural isomorphisms. We conclude that $\eta_X : X \to UR$ is a biuniversal arrow from $X$ to the 2-functor $U$. □

REMARK 10.8. Although it is not necessary, we can construct the factorizing 2-cell $\nu'$ on page 84 as follows. Let $H : X \to UD$ be a morphism of pseudo $S$-algebras. Then $\psi(H) = H'$ satisfies

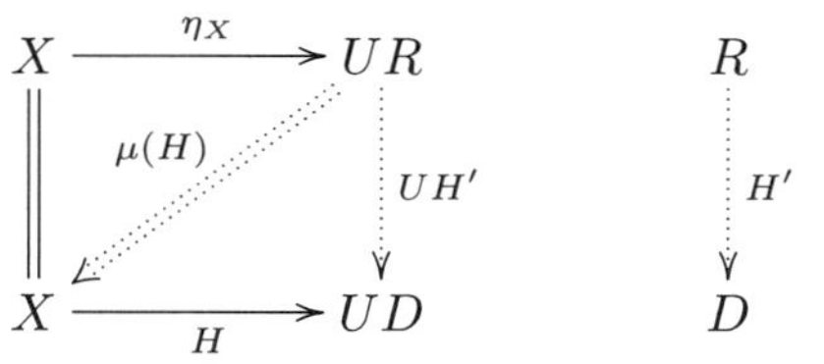

and $\mu(H)$ is the identity 2-cell. Suppose $\bar{H}' : R \to D$ is another morphism of pseudo $T$-algebras and $\nu$ is a 2-cell as follows.

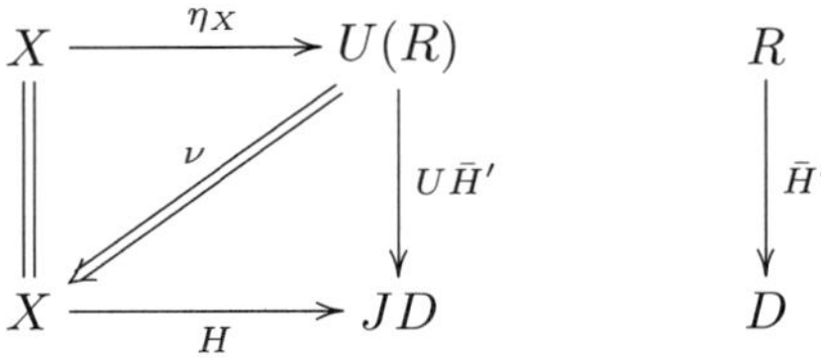

Define a 2-cell $\nu' : \bar{H}' \Rightarrow H'$ as follows. For $A \in Obj\ X \subseteq Obj\ R$, $\nu'A := \nu A$. If $w \in T'(n)$ and $\nu'$ is already defined for $A_1, \ldots, A_n \in Obj\ R$, then $\nu'(w(A_1, \ldots, A_n)) := \Phi(w)(\nu'A_1, \ldots, \nu'A_n) \circ \rho^{\bar{H}'}(A_1, \ldots, A_n)$. By induction $\nu'$ is a natural transformation. It also commutes with $\rho^{H'}$ and $\rho^{\bar{H}'}$ appropriately by construction. Hence $\nu'$ is a 2-cell in the 2-category of pseudo $T$-algebras. By construction we see that

(10.3)
$$\begin{array}{ccc} \bar{H}' & U\bar{H}' \circ \eta_X \xRightarrow{\nu} H \\ \Downarrow \nu' & \Downarrow U\nu' * i_{\eta_X} & \| \\ H' & UH' \circ \eta_X \xRightarrow{\mu(H)} H \end{array}$$

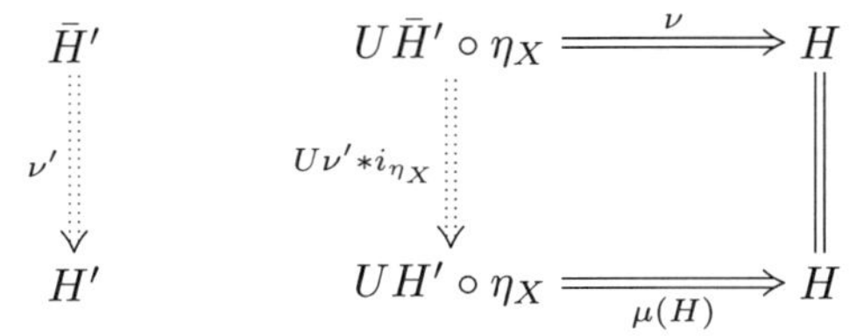

commutes. Such a 2-cell $\nu'$ is unique by the requirement that (10.3) commutes and by the commutivity with $\rho^{\bar{H}'}$ and $\rho^{H'}$ required of 2-cells $\bar{H}' \Rightarrow H'$. More precisely, the commutivity of (10.3) says that $\nu'A = \nu A$ for all $A \in Obj\ X$ and the appropriate commutivity with $\rho^{\bar{H}'}$ and $\rho^{H'}$ specifies what $\nu'$ does to objects of the form $w(A_1, \ldots, A_n)$ for $A_1, \ldots, A_n \in Obj\ R$. If $\nu$ is iso, then so is $\nu'$ by the construction and the fact that $\rho^{\bar{H}'}$ is iso.

THEOREM 10.9. *Let $S$ and $T$ be theories and $\phi : S \to T$ a morphism of theories. Then the forgetful 2-functor $U$ associated to $\phi$ from the 2-category of small pseudo $T$-algebras to the 2-category of small pseudo $S$-algebras admits a left biadjoint denoted $F$. Moreover, this pseudo functor $F$ is actually a strict 2-functor.*

*Proof:* For every pseudo $S$-algebra $X$ there exists a pseudo $T$-algebra $R$ and a biuniversal arrow $\eta_X : X \to UR$ by Lemma 10.7. This guarantees the existence of a left biadjoint by Theorem 9.17.

We can prove that $F$ is strict by inspecting its coherence isos constructed in the general theory of Theorem 9.17. Let $\mathcal{X}$ be the 2-category of pseudo $S$-algebras, let $\mathcal{A}$ be the 2-category of pseudo $T$-algebras, and let $G := U : \mathcal{A} \to \mathcal{X}$ be the forgetful 2-functor. For any pseudo $S$-algebra $X \in Obj\ \mathcal{X}$, we define $FX$ to be the free pseudo $T$-algebra $R$ on the pseudo $S$-algebra $X$ associated to the morphism of theories $\phi : S \to T$. The co-unit $\mu$ for the biuniversal $\eta_X : X \to UR$ is the identity as we observed in Lemma 10.7. The pseudo functor $U = G$ is actually a strict 2-functor, so $\delta^G$ and $\gamma^G$ are identity natural transformations. After inspecting diagram (9.17) on page 101, we see that $\delta^F_*$ must be trivial because $(\delta^G_{FX*})^{-1} * i_{\eta_X}$ and $\mu_{X,FX}(\eta_X \circ 1_X) = \mu(\eta_X \circ 1_X)$ are trivial. Hence $F$ preserves identities.

Similarly, each of the 2-cells in diagram (9.18) on page 101 is trivial, and therefore their composition is trivial. After inspecting diagram (9.19) on page 101, we see that $\gamma^F_{f,g}$ must also be trivial because both the horizontal top and bottom arrows are trivial. Therefore $F$ preserves compositions.

Since $F$ preserves compositions and identities, it is a strict 2-functor. □

THEOREM 10.10. *The biuniversal arrows $\eta_X : X \to UFX$ define a strict 2-natural transformation $\eta : 1_{\mathcal{X}} \Rightarrow U \circ F$, where $\mathcal{X}$ is the 2-category of pseudo S-algebras.*

*Proof:* Recall that the counits $\mu$ for the biuniversal arrows $\eta_X$ are all trivial as indicated on page 118 in Lemma 10.7. In the proof of Theorem 9.17 on page 109 the biuniversal arrows $\eta_X : X \to UFX$ are made into a pseudo natural transformation by defining $\tau_f := \mu_{X,FY}(\eta_Y \circ f)$ for $f : X \to Y$. We see that $\tau_f$ is trivial because $\mu_{X,FY}$ is trivial. Hence $\eta$ is strictly 2-natural. □

Theorem 10.9 can be sharpened. Let $\mathcal{A}$ denote the 2-category of pseudo $T$-algebras and let $\mathcal{X}$ denote the 2-category of pseudo $S$-algebras. Then the equivalence of categories $Mor_{\mathcal{A}}(FX, A) \to Mor_{\mathcal{X}}(X, UA)$ implicit in Theorem 10.9 is strictly 2-natural in each variable. However, it can be shown that a left 2-adjoint does not exist in specific cases. The equivalence in the other direction $Mor_{\mathcal{X}}(X, UA) \to Mor_{\mathcal{A}}(FX, A)$ in Theorem 10.9 is not strictly 2-natural in each variable. In fact, there is an example where there does not exist an equivalence $Mor_{\mathcal{X}}(X, UA) \to Mor_{\mathcal{A}}(FX, A)$ which is strictly 2-natural in each variable, even after replacing $F$ by another biadjoint $F'$. Counterexamples will be given after presenting Theorem 10.11, which is a sharper version of Theorem 10.9.

THEOREM 10.11. *Let $S$ and $T$ be theories. Let $U : \mathcal{A} \to \mathcal{X}$ be the forgetful 2-functor associated to a morphism $S \to T$ of theories. Let $F$ denote the left biadjoint to $U$ introduced in Theorem 10.9. Then the equivalence of categories $\phi_{X,A} : Mor_{\mathcal{A}}(FX, A) \to Mor_{\mathcal{X}}(X, UA)$ from Theorem 10.9 defined by $\phi_{X,A}(f) := Uf \circ \eta_X$ is strictly 2-natural in each variable.*

*Proof:* The universal arrow $\eta_X : X \to UFX$ is the inclusion morphism. The functor $\phi_{X,A} : Mor_{\mathcal{A}}(FX, A) \to Mor_{\mathcal{X}}(X, UA)$ is defined by $\phi_{X,A}(f) := Uf \circ \eta_X$ as in Lemma 9.13. The functor $\phi_{X,A}$ is an equivalence of categories for all $X \in Obj\ \mathcal{X}$ and all $A \in Obj\ \mathcal{A}$ because $\eta_X$ is a biuniversal arrow. The coherence isos $\tau'$ for the pseudo naturality of $\phi_{-,A}$ are defined on page 94 in terms of some trivial 2-cells, $\gamma^G$, and $\tilde{\tau}$, where $\tilde{\tau}$ is the coherence iso for $\eta$. But $\gamma^G$ is trivial for $G = U$ because $U$ is a strict 2-functor. The coherence iso $\tilde{\tau}$ is also trivial because $\eta$ is a strict 2-natural transformation. Hence $\tau'$ is also trivial and $\phi_{-,A}$ is strictly 2-natural, *i.e.* $\phi$ is 2-natural in the first variable.

The coherence isos $\tau$ for $\phi_{X,-}$ are defined on page 98 for morphisms $k : A \to A'$ by $\tau_{A,A'}(k) : e \mapsto \gamma^G_{e,k} * i_{\eta_X}$. But $G = U$ is a strict functor and $\gamma^G$ is trivial, hence $\tau$ is also trivial. Therefore $\phi_{X,-}$ is strictly 2-natural, *i.e.* $\phi$ is 2-natural in the second variable. We conclude that $X, A \mapsto \phi_{X,A}$ is strictly 2-natural in each variable. □

Before proving that Theorem 10.9 cannot be further improved to a left 2-adjoint, we need a theorem which states that we can change a morphism of pseudo $T$-algebras in a specific way and still have a morphism of pseudo $T$-algebras.

THEOREM 10.12. *Let $X, Y$ be pseudo $T$-algebras and $H : X \to Y$ a morphism of pseudo $T$-algebras. Suppose that $J_0(x) \in Obj\ Y$ and $\alpha_0(x) : J_0(x) \to H(x)$ is an isomorphism for each $x \in Obj\ X$. Then there exists a morphism $J : X \to Y$ of pseudo $T$-algebras whose object function is $J_0$ and there exists an iso 2-cell $\alpha$ :*

$J \Rightarrow H$ *of pseudo* $T$*-algebras such that* $\alpha(x) = \alpha_0(x)$ *for all* $x \in Obj\ X$*. Moreover, such* $J$ *and* $\alpha$ *are unique.*

*Proof:* For $x \in Obj\ X$ define $J(x) := J_0(x)$ and $\alpha(x) := \alpha_0(x)$. For a morphism $f : x_1 \to x_2$ of $X$ define $J(f) := \alpha(x_2)^{-1} \circ H(f) \circ \alpha(x_1)$. We easily see that $J$ is a functor and $\alpha$ is natural transformation from $J$ to the functor underlying $H$.

For $w \in T(n)$ let $\rho_w^H : H \circ \Phi(w) \Rightarrow \Psi \circ (H, \ldots, H)$ denote the coherence isomorphism for $H$, where $\Phi$ and $\Psi$ denote the structure maps of $X$ and $Y$ respectively. Define a natural isomorphism $\rho_w^J : J \circ \Phi(w) \Rightarrow \Psi \circ (J, \ldots, J)$ by the following diagram.

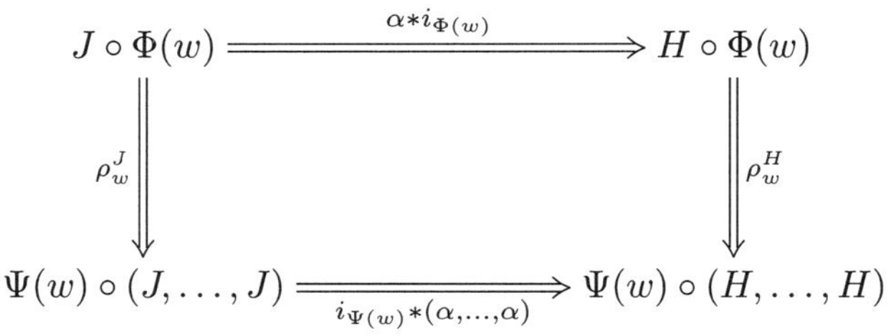


In other words $\rho_w^J := (i_{\Psi(w)} * (\alpha^{-1}, \ldots, \alpha^{-1})) \odot \rho_w^H \odot (\alpha * i_{\Phi(w)})$. This is a natural transformation because it consists of horizontal and vertical compositions of natural transformations.

We claim that $\rho_w^J$ satisfies the coherence diagrams required to make $J$ a morphism of pseudo $T$-algebras. We can prove the commutivity of any $J$ coherence diagram from the commutivity of the analogous $H$ coherence diagram by using the following procedure. First we draw the commutative $H$ coherence diagram and then we circumscribe it with the analogous $J$ coherence diagram. Next we draw the obvious isomorphisms between respective $J$ and $H$ vertices. All of the resulting inner diagrams commute because of the interchange law, because of the definition of $\rho_w^J$, or because of the diagram for $H$.

We present the substitution diagram to clarify the process. Let $f : \{1, \ldots, m\} \to \{1, \ldots, n\}$ be a function and $w \in T(m)$.

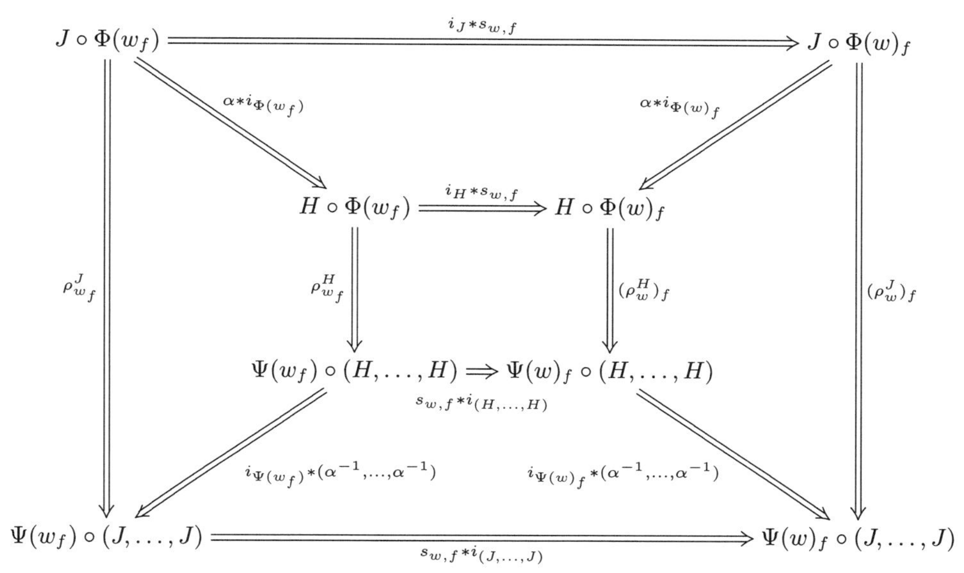


The top and bottom squares commute because of the interchange law. The left and right squares commute because of the definitions of $\rho^J_{w_f}$ and $\rho^J_w$. The innermost square commutes because $H$ is a morphism of pseudo $T$-algebras. Hence the outer rectangle commutes and $J$ satisfies the substitution coherence diagram.

The other diagrams can be verified using the same procedure. The only subtlety in this procedure occurs in the right hand vertical composition of the composition axiom. We reproduce the right hand part of that diagram obtained by the procedure mentioned above.

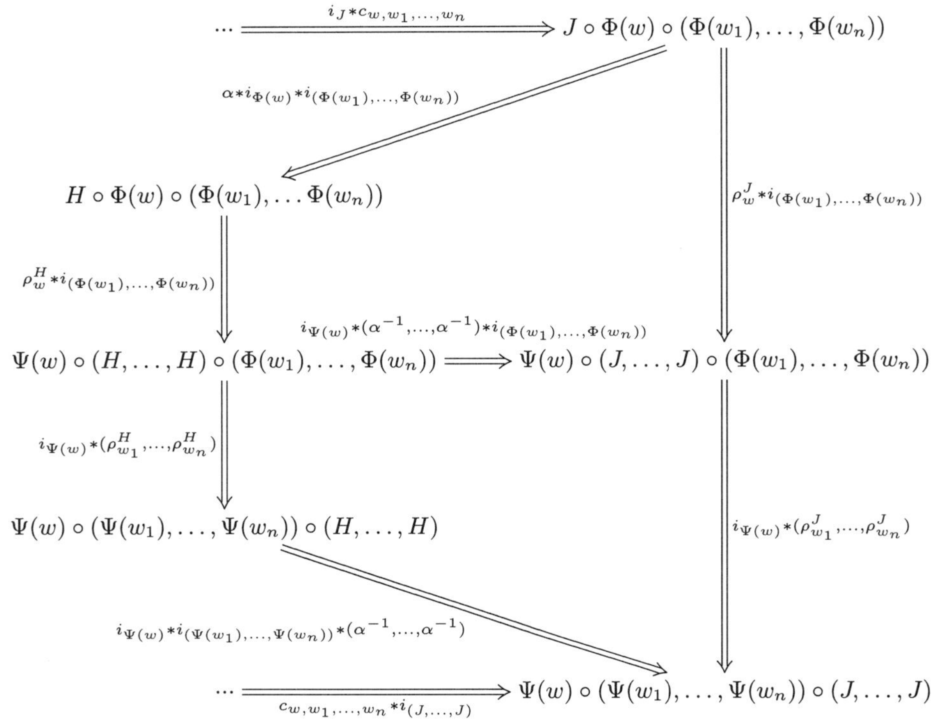


The upper right quadrilateral results from the diagram defining $\rho^J_w$ by horizontally composing with $i_{(\Phi(w_1),\dots,\Phi(w_n))}$. Then the upper right square commutes by iterated use of the interchange law.

The bottom right quadrilateral results from the defining diagrams of $\rho^J_{w_1}, \dots, \rho^J_{w_n}$ by taking their product, horizontally composing with the identity 2-cell

$$i_{(\Psi(w_1),\dots,\Psi(w_n))} = (i_{\Psi(w_1)}, \dots, i_{\Psi(w_n)}),$$

and finally reversing one of the arrows. The commutivity then follows from the interchange law.

The other parts of the diagram are easily seen to commute, and we conclude that $J$ satisfies the composition coherence.

The commutivity of all of these coherence diagrams implies that $J$ is a morphism of pseudo $T$-algebras. We conclude that $\alpha$ is a 2-cell in the 2-category of pseudo $T$-algebras by looking at its defining diagram.

Now we turn to the uniqueness. Suppose $J' : X \to Y$ is a morphism of pseudo $T$-algebras and $\alpha' : J' \Rightarrow H$ is a 2-cell in the 2-category of pseudo $T$-algebras such that for all $x \in Obj\ X$ we have $J'(x) = J_0(x)$ and $\alpha'(x) = \alpha_0(x)$. Then for a morphism $f : x_1 \to x_2$ in $X$ the diagram

$$\begin{array}{ccc} J_0(x_1) & \xrightarrow{\alpha(x_1)} & H(x_1) \\ {\scriptstyle J'(f)}\downarrow & & \downarrow{\scriptstyle H(f)} \\ J_0(x_1) & \xrightarrow[\alpha(x_2)]{} & H(x_2) \end{array}$$

commutes. Hence $J'(f) = \alpha(x_2)^{-1} \circ H(f) \circ \alpha(x_1) = J(f)$. For a word $w \in T(n)$, the diagram

$$\begin{array}{ccc} J' \circ \Phi(w) & \xRightarrow{\alpha * i_{\Phi(w)}} & H \circ \Phi(w) \\ {\scriptstyle \rho_w^{J'}}\Downarrow & & \Downarrow{\scriptstyle \rho_w^H} \\ \Psi(w) \circ (J', \ldots, J') & \xRightarrow[i_{\Psi(w)} * (\alpha, \ldots, \alpha)]{} & \Psi(w) \circ (H, \ldots, H) \end{array}$$

commutes. Hence $\rho_w^{J'} = (i_{\Psi(w)} * (\alpha^{-1}, \ldots, \alpha^{-1})) \odot \rho_w^H \odot (\alpha * i_{\Phi(w)}) = \rho_w^J$. We conclude $J' = J$ as morphisms of pseudo $T$-algebras. $\square$

LEMMA 10.13. *The functor* $\psi_{X,A} : Mor_{\mathcal{X}}(X, UA) \to Mor_{\mathcal{A}}(FX, A)$ *in Theorem 10.9 is not strictly 2-natural in each variable.*

*Proof:* Suppose $\psi$ is strictly 2-natural. Then for any morphism of pseudo $T$-algebras $J : FX \to FX$ the following diagram must commute.

$$\begin{array}{ccc} Mor_{\mathcal{A}}(FX, FX) & \xleftarrow{\psi_{X,FX}} & Mor_{\mathcal{X}}(X, UFX) \\ {\scriptstyle J_*}\downarrow & & \downarrow{\scriptstyle (UJ)_*} \\ Mor_{\mathcal{A}}(FX, FX) & \xleftarrow[\psi_{X,FX}]{} & Mor_{\mathcal{X}}(X, UFX) \end{array} \tag{10.4}$$

According to page 117, the output $\psi_{X,FX}(H)$ is always a strict morphism of pseudo $T$-algebras for all morphisms $H : X \to UFX$ of pseudo $S$-algebras. Let $a \in Obj\ FX$. Let $w$ be the trivial word in the theory $T$. Then $w(\psi_{X,FX}(\eta_X)(a))$ is isomorphic to (but not equal to) $\psi_{X,FX}(\eta_X)(a)$ via a coherence isomorphism. By Theorem 10.12 we can construct from this data a morphism $J : FX \to FX$ of pseudo $T$-algebras such that $J(w(\psi_{X,FX}(\eta_X)(a))) = \psi_{X,FX}(\eta_X)(a)$ and $J$ is the identity on all other objects. Chasing $\eta_X$ along diagram (10.4) from the top right corner, we see that $\psi_{X,FX}(UJ \circ \eta_X) = J \circ \psi_{X,FX}(\eta_X)$ and $J \circ \psi_{X,FX}(\eta_X)$ must be strict because $\psi_{X,FX}(UJ \circ \eta_X)$ is. But $J \circ \psi_{X,FX}(\eta_X)$ is not strict because it does not commute with the application of $w$ by the construction of $J$. $\square$

In fact, we present an example where there is no pseudo natural transformation $\psi$ as in Lemma 10.13 that is strictly 2-natural in the second variable, even after replacing $F$ by another left biadjoint to $U$. The reason is that our morphisms of pseudo algebras are not required to be strict, *i.e.* they are not required to commute with the structure maps.

EXAMPLE 10.14. Let $S$ be the trivial theory and let $T$ be the theory of commutative monoids. Let $\mathcal{X}$ be the 2-category of pseudo $S$-algebras and let $\mathcal{A}$ be the 2-category of pseudo $T$-algebras. Let $U : \mathcal{A} \to \mathcal{X}$ be the forgetful 2-functor associated to the trivial map of theories $S \to T$. Then there does not exist a left biadjoint $F' : \mathcal{X} \to \mathcal{A}$ which admits equivalences of categories $\psi'_{X,A} : Mor_{\mathcal{X}}(X, UA) \to Mor_{\mathcal{A}}(F'X, A)$ that are strictly 2-natural in the second variable.

*Proof:* First we prove that our constructed left biadjoint $F : \mathcal{X} \to \mathcal{A}$ does not admit equivalences $\psi'_{X,A}$ that are strictly 2-natural in the second variable. Suppose for each $X \in Obj\ \mathcal{X}$ there exist equivalences $\psi'_{X,A} : Mor_{\mathcal{X}}(X, UA) \to Mor_{\mathcal{A}}(FX, A)$ that are strictly natural in $A$, the second variable. Let $\phi'_{X,A}$ be a functor such that $\phi'_{X,A} \circ \psi'_{X,A}$ and $\psi'_{X,A} \circ \phi'_{X,A}$ are naturally isomorphic to the respective identities.

Let $X$ be the pseudo $S$-algebra with only one object $*$ and no nontrivial morphisms. Let $A$ be the category of finite sets with a choice of disjoint union. This makes $A$ into a pseudo $T$-algebra.

We claim that there exists a morphism $H : X \to UA$ of pseudo $S$-algebras such that $\psi'_{X,A}(H)(*) \neq \emptyset$. Suppose not. Then for every morphism $H : X \to UA$, we have $\psi'_{X,A}(H)(w(*, \ldots, *)) \cong w(\emptyset, \ldots, \emptyset) = \emptyset$ and thus $\psi'_{X,A}(H)$ is constant $\emptyset$. By the equivalence, every morphism $K : FX \to A$ of pseudo $T$-algebras is isomorphic to $\psi'_{X,A} \circ \phi'_{X,A}(K)$. This implies that $K$ must also be constant $\emptyset$. But this is a contradiction, since there are nontrivial morphisms $FX \to A$. Thus there exists a morphism $H : X \to UA$ of pseudo $S$-algebras such that $\psi'_{X,A}(H)(*) \neq \emptyset$.

We claim that there exists an object $x \in Obj\ FX$ such that $\psi'_{X,A}(H)(x) \neq H(*)$. Let $n \in \mathbb{N}$ be large enough that

$$n \cdot |\psi'_{X,A}(H)(*)| > |H(*)|.$$

This is possible because $|\psi'_{X,A}(H)(*)| \neq 0$ from above. Let $x = * + (* + (* + \cdots))$ where there are $n$ copies of $*$. Then $|\psi'_{X,A}(H)(x)| = n \cdot |\psi'_{X,A}(H)(*)|$ because $\psi'_{X,A}(H)$ is a morphism of pseudo $T$-algebras and isomorphisms in $A$ are bijections of sets. Thus $\psi'_{X,A}(H)(x) \neq H(*)$.

Let $J_0(\psi'_{X,A}(H)(x))$ be any set of the same cardinality as $\psi'_{X,A}(H)(x)$ but not equal to $\psi'_{X,A}(H)(x)$. Let $\alpha_0(\psi'_{X,A}(H)(x)) : J_0(\psi'_{X,A}(H)(x)) \to \psi'_{X,A}(H)(x)$ be a bijection. Let $J_0(a) = a$ for all $a \in Obj\ A$ such that $a \neq \psi'_{X,A}(H)(x)$. Then by Theorem 10.12 there exists a morphism $J : A \to A$ of pseudo $T$-algebras which is the identity except on the object $\psi'_{X,A}(H)(x)$. In particular $J(H(*)) = H(*)$ because $H(*) \neq \psi'_{X,A}(H)(x)$ from above.

The 2-naturality in the second variable implies that

$$\begin{array}{ccc} Mor_{\mathcal{A}}(FX, A) & \xleftarrow{\psi'_{X,A}} & Mor_{\mathcal{X}}(X, UA) \\ {\scriptstyle J_*}\downarrow & & \downarrow{\scriptstyle (UJ)_*} \\ Mor_{\mathcal{A}}(FX, A) & \xleftarrow[\psi'_{X,A}]{} & Mor_{\mathcal{X}}(X, UA) \end{array} \tag{10.5}$$

commutes, *i.e.* $J \circ \psi'_{X,A}(H) = \psi'_{X,A}(UJ \circ H)$. But $UJ \circ H = H$ because $J(H(*)) = H(*)$. Hence $J \circ \psi'_{X,A}(H) = \psi'_{X,A}(H)$. Evaluating this on $x$ gives

$$J(\psi'_{X,A}(H)(x)) = \psi'_{X,A}(H)(x)$$

which contradicts

$$J(\psi'_{X,A}(H)(x)) \neq \psi'_{X,A}(H)(x).$$

Thus there cannot exist such a $\psi'_{X,A} : Mor_{\mathcal{X}}(X, UA) \to Mor_{\mathcal{A}}(FX, A)$ and the reason is that we allow morphisms which are not strict.

Let $F' : \mathcal{X} \to \mathcal{A}$ be any left biadjoint for $U : \mathcal{A} \to \mathcal{X}$. Suppose it admits equivalences of categories $\psi'_{X,A} : Mor_{\mathcal{X}}(X, UA) \to Mor_{\mathcal{A}}(F'X, A)$ that are strictly 2-natural in the second variable. Since $F$ and $F'$ are left biadjoints for $U$, there exists for each $X$ a pseudo isomorphism $FX \to FX'$ by the biuniversal arrow argument in Lemma 9.7 and Theorem 9.20. This pseudo isomorphism induces an equivalence of categories $Mor_{\mathcal{A}}(F'X, A) \to Mor_{\mathcal{A}}(FX, A)$ which is strictly 2-natural in $A$. Composing this with $\psi'_{X,A}$ gives an equivalence of categories $Mor_{\mathcal{X}}(X, UA) \to Mor_{\mathcal{A}}(FX, A)$ which is strictly 2-natural in $A$, the second variable. But it was shown above that such a 2-natural equivalence cannot exist. Hence we have arrived at a contradiction and we conclude that $F'$ does not admit equivalences $\psi'_{X,A} : Mor_{\mathcal{X}}(X, UA) \to Mor_{\mathcal{A}}(F'X, A)$ that are strictly 2-natural in the second variable. □

We can build on the previous example to show that there does not exist a left 2-adjoint to the forgetful 2-functor in that situation.

EXAMPLE 10.15. Let $S$ be the trivial theory and let $T$ be the theory of commutative monoids. Let $\mathcal{X}$ be the 2-category of pseudo $S$-algebras and let $\mathcal{A}$ be the 2-category of pseudo $T$-algebras. Let $U : \mathcal{A} \to \mathcal{X}$ be the forgetful 2-functor associated to the trivial map of theories $S \to T$. Then there does not exist a left 2-adjoint to $U$, *i.e.* there does not exist a 2-functor $F' : \mathcal{X} \to \mathcal{A}$ which admits isomorphisms of categories $\phi_{X,A} : Mor_{\mathcal{A}}(F'X, A) \to Mor_{\mathcal{X}}(X, UA)$ that are strictly 2-natural in each variable.

*Proof:* Suppose such a $\phi$ existed. Let $\psi_{X,A} := \phi^{-1}_{X,A}$. Then $\psi_{X,A}$ is strictly 2-natural in the second variable $A$ and is an equivalence of categories. But this is impossible by the previous example. □

CHAPTER 11

# Weighted Bicolimits of Pseudo $T$-Algebras

In this chapter we show that the 2-category of pseudo $T$-algebras admits weighted bicolimits. The proof builds on the free pseudo $T$-algebra construction from Chapter 10 as well as the construction of pseudo colimits in the 2-category of small categories from Chapter 4. The present construction of bicolimits does not capture pseudo colimits because of the equivalence of morphism categories inherent to the construction of the free pseudo $T$-algebra. This equivalence arises because the morphisms of pseudo $T$-algebras are pseudo morphisms of pseudo $T$-algebras rather than strict morphisms. After proving that this 2-category admits bicolimits and bitensor products, we conclude that it admits weighted bicolimits.

THEOREM 11.1. *The 2-category $\mathcal{C}$ of small pseudo $T$-algebras admits bicolimits.*

*Proof:* Let $\mathcal{J}$ be a small 1-category and $F : \mathcal{J} \to \mathcal{C}$ a pseudo functor. In the following construction we use notation similar to the construction of the biuniversal arrows for forgetful 2-functors in Chapter 10.

First we define candidates $W \in Obj\ \mathcal{C}$ and $\pi : F \Rightarrow \Delta_W$. Let $T'$ denote the free theory on the sequence of sets $T(0), T(1), \ldots$ underlying the theory $T$. Let $Alg'$ be the category of small $T'$-algebras. Let $Graph'$ be the category of small directed graphs whose object sets are discrete $T'$ algebras. Then there is a forgetful functor $Alg' \to Graph'$ and it admits a left adjoint $V'$ by Freyd's Adjoint Functor Theorem.

Let $Obj\ R_{G'}$ be the free (discrete) $T'$ algebra on the set $\coprod_{j \in Obj\ \mathcal{J}} Obj\ Fj$. Let $Mor\ R_{G'}$ be the collection of the following arrows:

(1) For every $n \in \mathbf{N}$, for all words $w \in T(n)$, $w_1 \in T(m_1), \ldots, w_n \in T(m_n)$, and for all objects $A_1^1, \ldots, A_{m_1}^1, A_1^2, \ldots, A_{m_2}^2, \ldots, A_1^n, \ldots, A_{m_n}^n \in Obj\ R_{G'}$ there are arrows

$$c_{w,w_1,\ldots,w_n}(A_1^1, \ldots, A_{m_n}^n) :$$

$$w \circ (w_1, \ldots, w_n)(A_1^1, \ldots, A_{m_n}^n) \longrightarrow w(w_1(A_1^1, \ldots, A_{m_1}^1), \ldots, w_n(A_1^n, \ldots, A_{m_n}^n))$$

$$c^{-1}_{w,w_1,\ldots,w_n}(A_1^1, \ldots, A_{m_n}^n) :$$

$$w(w_1(A_1^1, \ldots, A_{m_1}^1), \ldots, w_n(A_1^n, \ldots, A_{m_n}^n)) \to w \circ (w_1, \ldots, w_n)(A_1^1, \ldots, A_{m_n}^n).$$

Here $w \circ (w_1, \ldots, w_n)$ is the composition in the original theory $T$. The target $w(w_1(A_1^1, \ldots, A_{m_1}^1), \ldots, w_n(A_1^n, \ldots, A_{m_n}^n))$ is the result of composing in the free theory and applying it to the $A$'s in the free algebra.

(2) For every $A \in Obj\ R_{G'}$ there are arrows

$$I_A : 1(A) \longrightarrow A$$

$$I_A^{-1} : A \longrightarrow 1(A).$$

Here 1 is the unit of the original theory $T$.

(3) For every word $w \in T(m)$, for every function $f : \{1, \ldots, m\} \to \{1, \ldots, n\}$, and for all objects $A_1, \ldots, A_n \in Obj\ R_{G'}$ there are arrows

$$s_{w,f}(A_1, \ldots, A_n) : w_f(A_1, \ldots, A_n) \longrightarrow w(A_{f1}, \ldots, A_{fm})$$

$$s_{w,f}^{-1}(A_1, \ldots, A_n) : w(A_{f1}, \ldots, A_{fm}) \longrightarrow w_f(A_1, \ldots, A_n).$$

The substituted word $w_f$ is the substituted word in the original theory $T$. The target $w(A_{f1}, \ldots, A_{fm})$ is the result of substituting in $w$ in the free theory and then evaluating on the $A$'s.

(4) For every word $w \in T(n)$, $j \in Obj\ \mathcal{J}$, and objects $A_1, \ldots, A_n$ of $Fj$ there are arrows

$$\rho_w^{\pi_j}(A_1, \ldots, A_n) : \Phi_j(w)(A_1, \ldots, A_n) \longrightarrow w(A_1, \ldots, A_n)$$

$$(\rho_w^{\pi_j})^{-1}(A_1, \ldots, A_n) : w(A_1, \ldots, A_n) \longrightarrow \Phi_j(w)(A_1, \ldots, A_n),$$

where $\Phi_j$ denotes the structure maps of the pseudo $T$-algebra $Fj$.

(5) Include all elements of $\coprod_{j \in \mathcal{J}} Mor\ Fj$ in $Mor\ R_{G'}$.

(6) For every morphism $f : i \to j$ of $\mathcal{J}$ and every $x \in Obj\ Fi$ we include arrows

$$h_{(x,f)} : x \longrightarrow a_f(x)$$

$$h_{(x,f)}^{-1} : a_f(x) \longrightarrow x$$

as in the proof of Theorem 4.2, where $a_f = Ff : Fi \to Fj$.

With these arrows, $R_{G'}$ is an object of *Graph*$'$. Now we apply the functor $V'$ to the directed graph $R_{G'}$ to get a category $R'$ which is a $T'$-algebra.

Let $K$ be the smallest congruence on the category $R'$ with the following properties:

(1) All of the relations necessary to make the coherence arrows (including $\rho_w^{\pi_j}$) into natural transformations belong to $K$. For example, if $A, B \in Obj\ R'$ and $f : A \to B$ is a morphism of $R'$,then the relation $I_A \circ f = 1(f) \circ I_B$ belongs to $K$.

(2) All of the relations necessary to make the coherence arrows (including $\rho_w^{\pi_j}$) into isos are in $K$. For example, for every $A \in Obj\ R'$ the relations $I_A \circ I_A^{-1} = 1_A$ and $I_A^{-1} \circ I_A = 1_A$ are in $K$.

(3) All of the relations for pseudo algebras listed in Definition 7.1 belong to $K$, where the objects range over the objects of $R'$.

(4) The original composition relations in each of the categories $Fj$ belong to $K$ for all $j \in Obj\ \mathcal{J}$.

(5) The coherence diagrams necessary to make the inclusion $\pi_j : Fj \to R'$ into a morphism of pseudo $T$-algebras belong to $K$. These diagrams are listed in Definition 7.4. Note that these coherence diagrams will involve the arrows $\rho_w^{\pi_j}(A_1, \ldots, A_n)$ for $w \in T(n)$.

(6) All of the relations in the proof of Theorem 4.2 are in $K$.

(7) If the relations $f_1 = g_1, \ldots, f_n = g_n$ are in $K$ and $w \in T'(n)$, then the relation $w(f_1, \ldots, f_n) = w(g_1, \ldots g_n)$ is also in $K$.

Next we mod out by the congruence $K$ in $R'$ and we get a pseudo $T$-algebra $R =: W \in Obj\ \mathcal{C}$.

We define a pseudo natural transformation $\pi : F \Rightarrow \Delta_W$ as follows. For $j \in Obj\ \mathcal{J}$, define $\pi_j : Fj \to W$ to be the inclusion functor. The functor $\pi_j$ is a morphism of pseudo $T$-algebras because of the relations we modded out by. Define

$\tau_{i,j}(f)_x : \pi_i(x) \longrightarrow \pi_j \circ a_f(x)$ by $\tau_{i,j}(f)_x := h_{(x,f)}$ as in the proof of Theorem of 4.2. Then $x \mapsto \tau_{i,j}(f)_x$ is a 2-cell $\pi_i \Rightarrow \pi_j \circ a_f$ in the 2-category of pseudo $T$-algebras because of the relations we modded out by and because of the work in the proof of Theorem 4.2. By an argument similar to Lemma 4.3 we conclude that $\pi : F \Rightarrow \Delta_W$ is a pseudo natural transformation. The candidate for the bicolimit of $F$ is $W \in Obj\ \mathcal{C}$ with the pseudo cone $\pi : F \Rightarrow \Delta_W$. This concludes the definition of the candidate for the bicolimit of $F$.

Let $V \in Obj\ \mathcal{C}$. Define the functor $\phi : Mor_{\mathcal{C}}(W, V) \longrightarrow PseudoCone(F, V)$ by $b \mapsto b \circ \pi$ as before. We need to see that $\phi$ is an equivalence of categories.

LEMMA 11.2. *There is a functor $\psi : PseudoCone(F, V) \longrightarrow Mor_{\mathcal{C}}(W, V)$.*

*Proof:* First we define $\psi$ on objects. Let $\pi' : F \Rightarrow \Delta_V$ be a pseudo natural transformation which is natural up to the coherence iso 2-cells $\tau'$. From $\pi'$ we get a map of sets

$$\coprod_{j \in Obj\ \mathcal{J}} Obj\ Fj \longrightarrow Obj\ V$$

which induces a map

$$d : Obj\ R_{G'} \longrightarrow Obj\ V$$

of discrete $T'$ algebras. Define $d$ on arrows of $R_{G'}$ as follows:

- $dg := \pi'_j g$ for all $g \in Mor\ Fj$ and all $j \in Obj\ \mathcal{J}$
- $dh_{(x,f)} := \tau'_{i,j}(f)_x$ and $dh^{-1}_{(x,f)} := (\tau'_{i,j}(f)_x)^{-1}$ for $f : i \longrightarrow j$ in $\mathcal{J}$ and $x \in Obj\ Fi$
- $d$ takes a coherence arrow in $R_{G'}$ to the analogous coherence iso in $V$
- $d(\rho_w^{\pi_j}) := \rho_w^{\pi'_j}$ where $\rho_w^{\pi'_j}$ is the coherence iso of the morphism $\pi'_j : Fj \longrightarrow V$ of pseudo $T$-algebras, and similarly $d((\rho_w^{\pi_j})^{-1}) := (\rho_w^{\pi'_j})^{-1}$.

This defines a morphism $d : R_{G'} \longrightarrow V$ of the category *Graph'*, where part of the structure of the $T'$-algebra $V$ is forgotten. The adjoint $Graph' \longrightarrow Alg'$ to the forgetful functor $Alg' \longrightarrow Graph'$ gives us a morphism $R' \longrightarrow V$, which we also denote by $d$. Furthermore, $d : R' \longrightarrow V$ preserves the relations in $K$. Hence $d$ induces a map $b : R \longrightarrow V$ on the quotient and $d$ is a morphism of pseudo $T$-algebras. Note that the coherence isos of $b$ are trivial. This is how we define $\psi$ on objects: $\psi(\pi') := b$.

Let $\sigma, \sigma' \in Obj\ PseudoCone(F, V)$ and let $\Xi : \sigma \rightsquigarrow \sigma'$ be a morphism in the category $PseudoCone(F, V)$. Then define a 2-cell $\psi(\Xi) : \psi(\sigma) \Rightarrow \psi(\sigma')$ by $\psi(\Xi)_x := \Xi_j(x)$ for $x \in Obj\ Fj$ and continue the definition inductively by

$$\psi(\Xi)_{w(x_1,\ldots,x_n)} := \Psi(w)(\psi(\Xi)_{x_1}, \ldots, \psi(\Xi)_{x_n}),$$

where $\Psi$ denotes the structure maps of the pseudo $T$-algebra $V$. Another inductive argument shows that this assignment preserves compositions and identities. □

LEMMA 11.3. *The functor $\phi \circ \psi : PseudoCone(F, V) \longrightarrow PseudoCone(F, V)$ is the identity functor.*

*Proof:* This is similar to Lemma 4.6. The only difference here is that we must prove that the coherence isos for the morphism $\pi'_j : Fj \longrightarrow V$ of pseudo $T$-algebras are the same as the coherence isos for $(\phi \circ \psi(\pi'))_j$. But this is true because the coherence isos of $\psi(\pi')$ are trivial. □

LEMMA 11.4. *The composite functor* $\psi \circ \phi : Mor_{\mathcal{C}}(W, V) \to Mor_{\mathcal{C}}(W, V)$ *is naturally isomorphic to the identity functor.*

*Proof:* We construct a natural isomorphism $\eta : 1_{Mor_{\mathcal{C}}(W,V)} \Rightarrow \psi \circ \phi$. Let $b \in Obj\ Mor_{\mathcal{C}}(W, V)$. We define $\eta_b =: \alpha$ inductively. For all $j \in Obj\ \mathcal{J}$ and all $x \in Obj\ Fj \subseteq Obj\ W$ we have $\psi \circ \phi(b)(x) = b(x)$. Define

$$\alpha_x : b(x) \to \psi \circ \phi(b)(x)$$

to be the identity for such $x$. For $w \in T(n)$ and $x_1, \ldots, x_n \in \coprod_{j \in Obj\ \mathcal{J}} Obj\ Fj$ define

$$\alpha_{w(x_1,\ldots,x_n)} := \rho^b_w(x_1, \ldots, x_n).$$

Now let $x_1, \ldots, x_n \in Obj\ W$ and $w \in T(n)$. Suppose $\alpha_{x_1}, \ldots, \alpha_{x_n}$ are already defined. Then define

$$\alpha_{w(x_1,\ldots,x_n)} : b(w(x_1, \ldots, x_n)) \to \psi \circ \phi(b)(w(x_1, \ldots, x_n))$$

to be the composition

$$\begin{array}{c} b(w(x_1, \ldots, x_n)) \\ \big\downarrow {\scriptstyle \rho^b_w(x_1,\ldots,x_n)} \\ \Psi(w)(bx_1, \ldots, bx_n) \\ \big\downarrow {\scriptstyle \Psi(w)(\alpha_{x_1},\ldots,\alpha_{x_n})} \\ \Psi(w)(\psi \circ \phi(b)x_1, \ldots, \psi \circ \phi(b)x_n). \end{array}$$

Then the assignment $x \mapsto \alpha_x$ is a 2-cell in the category of pseudo $T$-algebras because it is natural and commutes with the coherence isos of $b$ and $\psi \circ \phi(b)$ by an inductive argument (recall the coherence isos of $\psi \circ \phi(b)$ are trivial). An inductive argument also shows that $b \mapsto \eta_b$ is natural. □

LEMMA 11.5. *The functor* $\phi : Mor_{\mathcal{C}}(W, V) \to PseudoCone(F, V)$ *defined by* $b \mapsto b \circ \pi$ *is an equivalence of categories.*

*Proof:* This follows immediately from the previous two lemmas. □

LEMMA 11.6. *The object* $W \in Obj\ \mathcal{C}$ *and the pseudo cone* $\pi : F \Rightarrow \Delta_W$ *comprise a bicolimit of* $F$.

*Proof:* This follows immediately from the previous lemma. □

This completes the proof that the 2-category of small pseudo $T$-algebras admits bicolimits. □

LEMMA 11.7. *The 2-category* $\mathcal{C}$ *of pseudo* $T$*-algebras admits bitensor products.*

*Proof:* Let $J$ be a category and $F$ a pseudo $T$-algebra. First we define an object $R_{G'}$ of $Graph'$. Let $Obj\ R_{G'}$ be the free discrete $T'$-algebra on the set $Obj\ J \times Obj\ F$, where $T'$ is the free theory on $T$. Let $Mor\ R_{G'}$ be the collection of the following arrows.

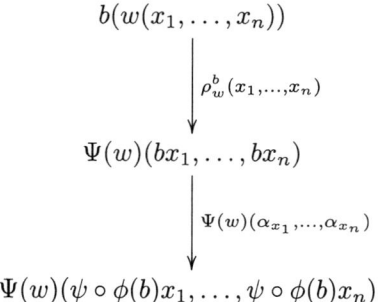

(1) For every $n \in \mathbf{N}$, for all words $w \in T(n)$, $w_1 \in T(m_1), \ldots, w_n \in T(m_n)$, and for all objects $A_1^1, \ldots, A_{m_1}^1, A_1^2, \ldots, A_{m_2}^2, \ldots, A_1^n, \ldots, A_{m_n}^n \in Obj\ R_{G'}$ there are arrows

$$c_{w,w_1,\ldots,w_n}(A_1^1, \ldots, A_{m_n}^n) :$$

$$w \circ (w_1, \ldots, w_n)(A_1^1, \ldots, A_{m_n}^n) \longrightarrow w(w_1(A_1^1, \ldots, A_{m_1}^1), \ldots, w_n(A_1^n, \ldots, A_{m_n}^n))$$

$$c_{w,w_1,\ldots,w_n}^{-1}(A_1^1, \ldots, A_{m_n}^n) :$$

$$w(w_1(A_1^1, \ldots, A_{m_1}^1), \ldots, w_n(A_1^n, \ldots, A_{m_n}^n)) \longrightarrow w \circ (w_1, \ldots, w_n)(A_1^1, \ldots, A_{m_n}^n).$$

Here $w \circ (w_1, \ldots, w_n)$ is the composition in the original theory $T$. The target $w(w_1(A_1^1, \ldots, A_{m_1}^1), \ldots, w_n(A_1^n, \ldots, A_{m_n}^n))$ is the result of composing in the free theory and applying it to the $A$'s in the free algebra.

(2) For every $A \in Obj\ R_{G'}$ there are arrows

$$I_A : 1(A) \longrightarrow A$$

$$I_A^{-1} : A \longrightarrow 1(A).$$

Here 1 is the unit of the original theory $T$.

(3) For every word $w \in T(m)$, for every function $f : \{1, \ldots, m\} \to \{1, \ldots, n\}$, and for all objects $A_1, \ldots, A_n \in Obj\ R_{G'}$ there are arrows

$$s_{w,f}(A_1, \ldots, A_n) : w_f(A_1, \ldots, A_n) \longrightarrow w(A_{f1}, \ldots, A_{fm})$$

$$s_{w,f}^{-1}(A_1, \ldots, A_n) : w(A_{f1}, \ldots, A_{fm}) \longrightarrow w_f(A_1, \ldots, A_n).$$

The substituted word $w_f$ is the substituted word in the original theory $T$. The target $w(A_{f1}, \ldots, A_{fm})$ is the result of substituting in $w$ in the free theory and then evaluating on the $A$'s.

(4) For every word $w \in T(n)$, $j \in Obj\ J$, and objects $x_1, \ldots, x_n$ of $F$ there are arrows

$$\rho_w^{\pi(j)}((j, x_1), \ldots, (j, x_n)) : (j, \Phi(w)(x_1, \ldots, x_n)) \longrightarrow w((j, x_1), \ldots, (j, x_n))$$

$$(\rho_w^{\pi(j)})^{-1} : w((j, x_1), \ldots, (j, x_n)) \longrightarrow (j, \Phi(w)(x_1, \ldots, x_n)),$$

where $\Phi$ denotes structure maps of the pseudo $T$-algebra $F$.

(5) Include all elements of $Mor\ J \times Mor\ F$ in $Mor\ R_{G'}$.

With these arrows, $R_{G'}$ is an object of $Graph'$. Now we apply the free $T'$-algebra functor to the directed graph $R_{G'}$ to get a category $R'$ which is a $T'$ algebra. Let $K$ be the smallest congruence on the category $R'$ with the following properties:

(1) All of the relations necessary to make the coherence arrows (including $\rho_w^{\pi(j)}$) into natural transformations belong to $K$. For example, if $A, B \in Obj\ R'$ and $f : A \to B$ is a morphism in $R'$,then the relation $I_A \circ f = 1(f) \circ I_B$ belongs to $K$.

(2) All of the relations necessary to make the coherence arrows (including $\rho_w^{\pi(j)}$) into isos are in $K$. For example, for every $A \in Obj\ R'$ the relations $I_A \circ I_A^{-1} = 1_A$ and $I_A^{-1} \circ I_A = 1_A$ are in $K$.

(3) All of the relations for pseudo algebras listed in Definition 7.1 belong to $K$, where the objects range over the objects of $R'$.

(4) The original composition relations in the category $J \times F$ belong to $K$.

(5) For each $j \in J$, the coherence diagrams necessary to make the inclusion $F \to R'$, $x \mapsto (j,x)$ into a morphism of pseudo $T$-algebras belong to $K$. These diagrams are listed in Definition 7.4. Note that these coherences will involve the arrows

$$\rho_w^{\pi(j)}((j,x_1),\ldots,(j,x_n)) : (j,\Phi(w)(x_1,\ldots,x_n)) \to w((j,x_1),\ldots,(j,x_n)).$$

(6) For any $g : j_1 \to j_2$ in $J$ and $x_1,\ldots,x_n$ in $F$ we include the relation

$$\begin{array}{ccc} (j_1,\Phi(w)(x_1,\ldots,x_n)) & \xrightarrow{(g,1_{\Phi(w)(x_1,\ldots,x_n)})} & (j_2,\Phi(w)(x_1,\ldots,x_n)) \\ {\scriptstyle \rho_w^{\pi(j_1)}(x_1,\ldots,x_n)}\downarrow & & \downarrow{\scriptstyle \rho_w^{\pi(j_2)}(x_1,\ldots,x_n)} \\ w((j_1,x_1),\ldots,(j_1,x_n)) & \xrightarrow[w((g,x_1),\ldots,(g,x_n))]{} & w((j_2,x_1),\ldots,(j_2,x_n)). \end{array}$$

(7) If the relations $f_1 = g_1,\ldots,f_n = g_n$ are in $K$ and $w \in T'(n)$, then the relation $w(f_1,\ldots,f_n) = w(g_1,\ldots g_n)$ is also in $K$.

Next we mod out by the congruence $K$ in $R'$ and we get a pseudo $T$-algebra $J * F \in Obj\ \mathcal{C}$. We define a functor $\pi : J \to \mathcal{C}(F, J*F)$ by

$$\pi(j)(x) := (j,x)$$
$$\pi(j)(f) := (1_j, f)$$
$$(\pi(g))_x := (g, 1_x)$$

for $j \in Obj\ J, x \in Obj\ F, f \in Mor\ F$, and $g \in Mor\ J$. Then $\pi(j) : F \to J * F$ is a morphism of pseudo $T$-algebras with coherence isos $\rho^{\pi(j)}$ and $\pi(g) : \pi(j_1) \Rightarrow \pi(j_2)$ is a 2-cell in the 2-category of pseudo $T$-algebras because of the relations. The relations also imply that $\pi$ is a functor.

We claim that $\pi$ induces an equivalence

$$\mathcal{C}(J*F, C) \xrightarrow{\phi} Cat(J, \mathcal{C}(F,C))$$
$$b \mapsto \mathcal{C}(F,b)\circ\pi$$
$$\alpha \mapsto \mathcal{C}(F,\alpha) * i_\pi$$

of categories. Define a functor $\psi : Cat(J,\mathcal{C}(F,C)) \to \mathcal{C}(J*F,C)$ as follows. For a functor $\sigma : J \to \mathcal{C}(F,C)$, we have a map of sets

$$Obj\ J \times Obj\ F \to Obj\ C$$
$$(j,x) \mapsto \sigma(j)(x)$$

which induces a map $\psi(\sigma) : Obj\ R_{G'} \to Obj\ C$ of discrete $T'$-algebras satisfying

$$\psi(\sigma)(j,x) := \sigma(j)(x)$$
$$\psi(\sigma)(w((j_1,x_1),\ldots,(j_n,x_n))) := \Phi^C(w)(\sigma(j_1)(x_1),\ldots,\sigma(j_n)(x_n))$$

for $(j,x),(j_1,x_1),\ldots,(j_n,x_n) \in J \times F$. Define $\psi(\sigma)$ on arrows of $R_{G'}$ by

$$\psi(\sigma)(c_{w,w_1,\ldots,w_n}(A_1^1,\ldots,A_{m_n}^n)) := c_{w,w_1,\ldots,w_n}(\psi(\sigma)(A_1^1),\ldots,\psi(\sigma)(A_{m_n}^n))$$
$$\psi(\sigma)(I_A) := I_{\psi(\sigma)(I_A)}$$
$$\psi(\sigma)(s_{w,f}(A_1,\ldots,A_n)) := s_{w,f}(\psi(\sigma)(A_1),\ldots,\psi(\sigma)(A_n))$$
$$\psi(\sigma)(g,f) := \sigma(j_2)(f)\circ\sigma(g)_{x_1} = \sigma(g)_{x_2}\circ\sigma(j_1)(f)$$

for $A^k_\ell, A, A_i \in Obj\ R_{G'}$, $f : m \to n$, $g : j_1 \to j_2$ in $J$, and $f : x_1 \to x_2$ in $F$. We define $\psi(\sigma)$ similarly for $c^{-1}_{w,w_1,\ldots,w_n}, I^{-1}_A, s^{-1}_{w,f}$. Then $\psi(\sigma) : R_{G'} \to C$ is a morphism in $Graph'$, which induces a morphism $R' \to C$ in $Alg'$. It preserves the relations and therefore induces a morphism $\psi(\sigma) : J * F \to C$ of pseudo $T$-algebras on the quotient. This is actually a strict morphism of pseudo $T$-algebras. For a natural transformation $\Xi : \sigma \Rightarrow \sigma'$ define a 2-cell $\psi(\Xi) : \psi(\sigma) \Rightarrow \psi(\sigma')$ inductively by

$$\psi(\Xi)_{(j,x)} := (\Xi_j)_x$$

for $(j,x) \in Obj\ J \times Obj\ F$ and

$$\psi(\Xi)_{w(A_1,\ldots,A_n)} := \Phi^C(w)(\psi(\Xi)_{A_1}, \ldots, \psi(\Xi)_{A_n})$$

whenever $\psi(\Xi)_{A_1}, \ldots, \psi(\Xi)_{A_n}$ are already defined. From these definitions we can conclude that $\psi$ is a functor and $\phi \circ \psi = 1_{Cat(J,\mathcal{C}(F,C))}$. For example,

$$\begin{aligned}(\phi \circ \psi(\sigma))(j)(x) &= (\psi(\sigma) \circ \pi(j))(x)\\ &= \psi(\sigma)(j,x)\\ &= \sigma(j)(x)\end{aligned}$$

and also

$$\begin{aligned}((\phi \circ \psi(\Xi))_j)_x &= ((\psi(\Xi) * i_\pi)_j)_x\\ &= \psi(\Xi)_{\pi(j)(x)}\\ &= (\Xi_j)_x.\end{aligned}$$

We construct a natural isomorphism $\eta : 1_{\mathcal{C}(J*F,C)} \Rightarrow \psi \circ \phi$. Let $b : J * F \to C$ be a morphism of pseudo $T$-algebras. We define $\eta_b =: \alpha$ inductively. For all $(j,x) \in Obj\ J \times Obj\ F$ we have

$$\begin{aligned}\psi \circ \phi(b)(j,x) &= \psi(\mathcal{C}(F,b) \circ \pi)(j,x)\\ &= (\mathcal{C}(F,b) \circ \pi)(j)(x)\\ &= (b \circ \pi(j))(x)\\ &= b(j,x).\end{aligned}$$

Define

$$\alpha_{(j,x)} : b(j,x) \to \psi \circ \phi(b)(j,x)$$

to be the identity for such $(j,x)$. For $w \in T(n)$ and $(j_1,x_1), \ldots, (j_n,x_n) \in Obj\ J \times Obj\ F$ define

$$\alpha_{w((j_1,x_1),\ldots,(j_n,x_n))} := \rho^b_w((j_1,x_1), \ldots, (j_n,x_n)).$$

For $A_1, \ldots, A_n \in Obj\ R_{G'} = Obj\ J * F$ and $w \in T(n)$, define

$$\alpha_{w(A_1,\ldots,A_n)} : b(w(A_1,\ldots,A_n)) \to \psi \circ \phi(b)(w(A_1,\ldots,A_n))$$

to be the composition

$$\begin{array}{c} b(w(A_1,\ldots,A_n)) \\ \Big\downarrow {\scriptstyle \rho^b_w(A_1,\ldots,A_n)} \\ \Psi(w)(bA_1,\ldots,bA_n) \\ \Big\downarrow {\scriptstyle \Psi(w)(\alpha_{A_1},\ldots,\alpha_{A_n})} \\ \Psi(w)(\psi\circ\phi(b)A_1,\ldots,\psi\circ\phi(b)A_n). \end{array}$$

Then the assignment $x \mapsto \alpha_x$ is a 2-cell in the category of pseudo $T$-algebras because it is natural and commutes with the coherence isos of $b$ and $\psi \circ \phi(b)$ by an inductive argument (recall the coherence isos of $\psi \circ \phi(b)$ are trivial). An inductive argument also shows that $b \mapsto \eta_b$ is natural.

By Remark 3.26, this implies that $J * F$ is a bitensor product of $J$ and $F$. □

THEOREM 11.8. *The 2-category $\mathcal{C}$ of pseudo $T$-algebras admits weighted bicolimits.*

*Proof:* The 2-category $\mathcal{C}$ admits bicoproducts and bicoequalizers by Theorem 11.1. It admits bitensor products by the previous lemma. Hence by Theorem 3.27 it admits weighted bicolimits. □

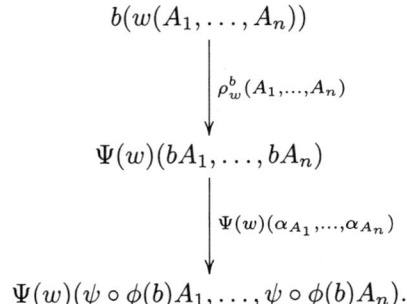

CHAPTER 12

# Stacks

In this chapter we introduce the language of stacks in analogy to sheaves, since stacks generalize sheaves. A stack is a contravariant pseudo functor from a Grothendieck topology to a 2-category which takes Grothendieck covers to bilimits in the sense described below. The target 2-category is required to admit bilimits. We have shown that the 2-category of pseudo algebras over a theory admits bilimits, so we can speak of stacks of pseudo algebras. Some references for stacks are [**13**], [**17**], [**18**], [**42**], and [**52**]. We are interested in stacks because we want to capture the algebraic structure of holomorphic families of rigged surfaces as in Section 13.3.

DEFINITION 12.1. A *basis for a Grothendieck topology* on a category $\mathcal{B}$ with pullbacks is a function $K$ which assigns to each object $B$ of $\mathcal{B}$ a collection of families of morphisms with codomain $B$ such that:

(1) If $g : B' \to B$ is an isomorphism, then $\{g\} \in K(B)$.
(2) If $\{g_i : B_i \to B | i \in I\} \in K(B)$, then for any morphism $g : D \to B$ the family $\{\pi_i^2 : B_i \times_B D \to D | i \in I\}$ of pullbacks of the $g_i$ along $g$ is in $K(D)$.
(3) If $\{g_i : B_i \to B | i \in I\} \in K(B)$ and $\{f_{ij} : D_{ij} \to B_i | j \in J_i\} \in K(B_i)$ for all $i$, then the composite family $\{g_i \circ f_{ij} : B_{ij} \to B | i \in I, j \in J_i\}$ is in $K(B)$.

The second axiom is called the *stability axiom* because it says that $K$ is stable under pullbacks. The third axiom is called the *transitivity axiom.* Often we refer to the basis as well as the category $\mathcal{B}$ as a Grothendieck topology. We follow this convention. Some authors call a Grothendieck topology a Grothendieck site. The elements of $K(B)$ are called *Grothendieck covers.*

DEFINITION 12.2. Let $\mathcal{B}$ be a Grothendieck topology and $\mathcal{C}$ a concrete category. Then a $\mathcal{C}$*-sheaf* on $\mathcal{B}$ is a contravariant functor $G : \mathcal{B} \to \mathcal{C}$ which takes Grothendieck covers to limits, *i.e.* for any object $B$ of $\mathcal{B}$ and for any Grothendieck cover $\{g_i : B_i \to B | i \in I\} \in K(B)$ the following diagram is an equalizer,

$$G(B) \xrightarrow{\ e\ } \prod_{i\in I} G(B_i) \overset{p_1}{\underset{p_2}{\rightrightarrows}} \prod_{i,j\in I} G(B_i \times_B B_j) \tag{12.1}$$

where $e(a) = \{G(g_i)a\}_{i\in I}$ and $p_1(\{a_k\}_{k\in I})_{ij} = G(\pi^1_{ij})a_i$ and $p_2(\{a_k\}_{k\in I})_{ij} = G(\pi^2_{ij})a_j$. Here $\pi^1_{ij}, \pi^2_{ij}$ are the morphisms in the pullback diagrams for $B_{ij} :=$

$B_i \times_B B_j$.

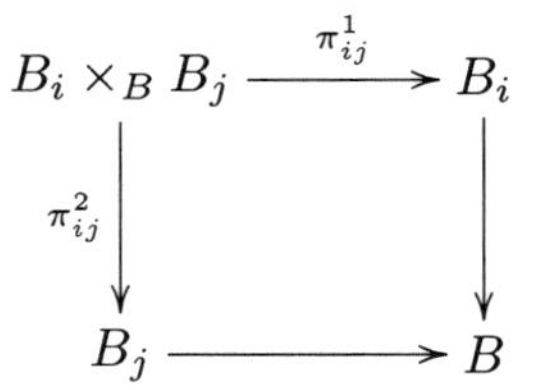


See [**40**] for a thorough discussion of Grothendieck topologies and sheaves. Diagram (12.1) is an equalizer if and only if it is *exact*. Usually we speak of a $\mathcal{C}$-sheaf as a sheaf of objects of $\mathcal{C}$. For example, if $\mathcal{C}$ is the category of sets, then we speak of a sheaf of sets. Next we speak of stacks of categories and then generalize to stacks of objects with algebraic structure.

Let $Cat$ denote the 2-category of small categories. Suppose $\mathcal{B}$ is a Grothendieck topology. Let $G : \mathcal{B} \to Cat$ be a contravariant pseudo functor. Let $B$ be an object of $\mathcal{B}$ and $\{g_i : B_i \to B | i \in I\} \in K(B)$ a Grothendieck cover. Consider the diagram (12.2)

$$\prod_{i \in I} G(B_i) \underset{p_2}{\overset{p_1}{\rightrightarrows}} \prod_{i,j \in I} G(B_i \times_B B_j) \underset{p_{23}}{\overset{p_{12}}{\underset{p_{13}}{\rightrightarrows}}}\!\!\!\!\to \prod_{i,j,k \in I} G(B_i \times_B B_j \times_B B_k)$$

where the arrows are defined as

$$p_1(\{a_k\}_k)_{ij} := G(\pi^1_{ij})a_i$$

$$p_2(\{a_k\}_k)_{ij} := G(\pi^2_{ij})a_j$$

$$p_{12}(\{a_{\ell m}\}_{\ell m})_{ijk} := G(\pi^{12}_{ijk})a_{ij}$$

$$p_{13}(\{a_{\ell m}\}_{\ell m})_{ijk} := G(\pi^{13}_{ijk})a_{ik}$$

$$p_{23}(\{a_{\ell m}\}_{\ell m})_{ijk} := G(\pi^{23}_{ijk})a_{jk}.$$

Here $\pi^{12}_{ijk}, \pi^{13}_{ijk}, \pi^{23}_{ijk}$ are the morphisms for the triple fiber product $B_i \times_B B_j \times_B B_k$ as in the following commutative diagram from [**52**]. The unlabelled arrows are $g_i, g_j$, and $g_k$ from the Grothendieck cover.

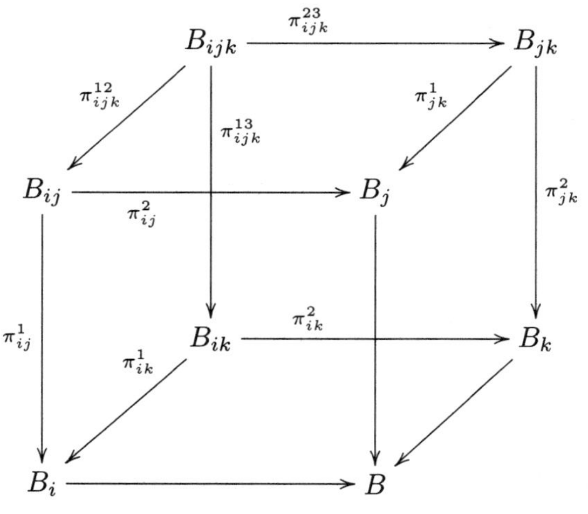

Every face in this diagram is a pullback square. The object $B_{ijk}$ is the limit of the diagram obtained from this one by deleting $B_{ijk}$ and the arrows emanating from it.

Diagram (12.2) can be interpreted as the image of a pseudo functor $F : \mathcal{J} \to Cat$ as follows. Let $\mathcal{J}$ be the free 1-category on the directed graph

$$X \underset{f_2}{\overset{f_1}{\rightrightarrows}} Y \overset{f_{12}}{\underset{f_{23}}{\overset{f_{13}}{\rightrightarrows\!\!\!\to}}} Z \tag{12.3}$$

modded out by the relations below.

$$\begin{array}{ccc} X & \xrightarrow{f_1} & Y \\ {\scriptstyle f_1}\downarrow & & \downarrow{\scriptstyle f_{12}} \\ Y & \xrightarrow[f_{13}]{} & Z \end{array} \qquad\qquad \begin{array}{ccc} X & \xrightarrow{f_2} & Y \\ {\scriptstyle f_2}\downarrow & & \downarrow{\scriptstyle f_{13}} \\ Y & \xrightarrow[f_{23}]{} & Z \end{array}$$

$$\begin{array}{ccc} X & \xrightarrow{f_1} & Y \\ {\scriptstyle f_2}\downarrow & & \downarrow{\scriptstyle f_{23}} \\ Y & \xrightarrow[f_{12}]{} & Z \end{array}$$

Define a covariant pseudo functor $F : \mathcal{J} \to Cat$ which takes diagram (12.3) to diagram (12.2) and takes identity morphisms to identity morphisms. The pseudo functor $F$ is defined on all possible composites of nontrivial morphisms as:

$$F(f_{12} \circ f_1)(\{a_\ell\}_\ell)_{ijk} := G(\pi^1_{ij} \circ \pi^{12}_{ijk})a_i$$
$$F(f_{13} \circ f_2)(\{a_\ell\}_\ell)_{ijk} := G(\pi^2_{ik} \circ \pi^{13}_{ijk})a_k$$
$$F(f_{23} \circ f_1)(\{a_\ell\}_\ell)_{ijk} := G(\pi^1_{jk} \circ \pi^{23}_{ijk})a_j.$$

The identity coherence isos $\delta^F$ for $F$ are equalities because $F$ takes identity morphisms to identity morphisms. The coherence isos $\gamma^F$ for composites of non-identity morphisms are defined as tuples of the composition coherence isos for $G$. For example, the coherence iso $\gamma^F_{f_1,f_{12}}\{a_\ell\}_\ell : F(f_{12}) \circ F(f_1)\{a_\ell\}_\ell \to F(f_{12} \circ f_1)\{a_\ell\}_\ell$ is defined as

$$\{\gamma^G_{\pi^{12}_{ijk},\pi^1_{ij}} a_i\}_{ijk} : \{G(\pi^{12}_{ijk}) \circ G(\pi^1_{ij})a_i\}_{ijk} \to \{G(\pi^1_{ij} \circ \pi^{12}_{ijk})a_i\}_{ijk}.$$

The coherence isos $\gamma^F$ for composites involving one or more identity morphisms are defined to be equalities. For example, the coherence iso

$$\gamma^F_{1_X,f_1}\{a_\ell\}_\ell : F(f_1) \circ F(1_X)\{a_\ell\}_\ell \to F(f_1 \circ 1_X)\{a_\ell\}_\ell$$

is equality. The coherence diagram in the pseudo functor unit axiom for $\delta^F$ is satisfied because of this definition. The coherence diagram in the pseudo functor composition axiom for $\gamma^F$ is satisfied because of the diagrams for $\gamma^G$ and also because of this definition. The coherence isos are also natural because $\mathcal{J}$ has no nontrivial 2-cells. Thus $F : \mathcal{J} \to Cat$ is a pseudo functor whose image is diagram (12.2). By a *bilimit of diagram (12.2)* we mean a bilimit of this functor $F$.

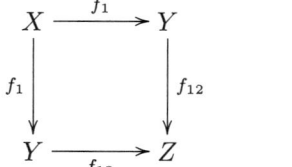

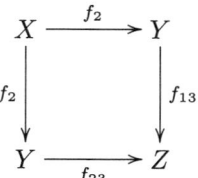

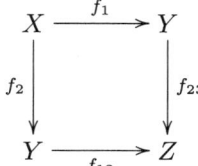

In the context of stacks there is a canonical candidate for the bilimit of $F$, namely $G(B)$. The candidate for the universal pseudo cone $\pi' : \Delta_{G(B)} \Rightarrow F$ is defined on objects as follows.

$$\pi'_X : G(B) \to \prod_i G(B_i)$$

$$\pi'_X(a) := \{G(g_i)a\}_i$$

$$\pi'_Y : G(B) \to \prod_{i,j} G(B_i \times_B B_j)$$

$$\pi'_Y(a) := \{G(g_i \circ \pi^1_{ij})a\}_{ij}$$

$$\pi'_Z : G(B) \to \prod_{i,j,k} G(B_i \times_B B_j \times_B B_k)$$

$$\pi'_Z(a) := \{G(g_i \circ \pi^1_{ij} \circ \pi^{12}_{ijk})a\}_{ijk}$$

The coherence isos $\tau'_f : Ff \circ \pi'_{Sf} \Rightarrow \pi'_{Tf} \circ \Delta_{G(B)}(f)$ for the pseudo cone $\pi'$ and non-identity morphisms $f$ in $\mathcal{J}$ are defined in terms of $\gamma^G$. For example, for $f_1 : X \to Y$ we have

$$\begin{array}{ccc} G(B) & \xrightarrow{\pi'_X} & \prod_i G(B_i) \\ {\scriptstyle 1_{G(B)} = \Delta_{G(B)}(f_1)} \downarrow & \swArrow{\scriptstyle \tau'_{f_1}} & \downarrow {\scriptstyle F(f_1) = p_1} \\ G(B) & \xrightarrow[\pi'_Y]{} & \prod_{i,j} G(B_i \times_B B_j) \end{array}$$

defined by $\tau'_{f_1} a := \{\gamma^G_{\pi^1_{ij}, g_i} a\}_{ij} : \{G(\pi^1_{ij}) \circ G(g_i)a\}_{ij} \to \{G(g_i \circ \pi^1_{ij})a\}_{ij}$ for all objects $a$ of $G(B)$. For the identity morphisms $1_X, 1_Y$, and $1_Z$ of $\mathcal{J}$ we define $\tau'_{1_X}, \tau'_{1_Y}$, and $\tau'_{1_Z}$ to be equalities. The coherence diagram for the unit axiom of pseudo natural transformations is satisfied because of this definition. The composition axiom for $\tau'$ and nontrivial morphisms is satisfied because of the composition axiom for $\gamma^G$ and because $\gamma^{\Delta_{G(B)}}$ is an equality. The composition axiom for $\tau'$ whenever one or more of the morphisms is trivial follows trivially. Thus $\pi' : \Delta_{G(B)} \Rightarrow F$ is a pseudo natural transformation with coherence isos $\tau'$. After these preliminary remarks, we can finally define stack of categories.

DEFINITION 12.3. Let $Cat$ denote the 2-category of small categories. Suppose $\mathcal{B}$ is a Grothendieck topology. A *stack of categories* is a contravariant pseudo functor $G : \mathcal{B} \to Cat$ which takes Grothendieck covers to bilimits, *i.e.* for any object $B$ of $\mathcal{B}$ and any Grothendieck cover $\{g_i : B_i \to B | i \in I\} \in K(B)$ the diagram

$$\prod_{i \in I} G(B_i) \overset{p_1}{\underset{p_2}{\rightrightarrows}} \prod_{i,j \in I} G(B_i \times_B B_j) \overset{p_{12}}{\underset{p_{23}}{\overset{p_{13}}{\rightrightarrows}}} \prod_{i,j,k \in I} G(B_i \times_B B_j \times_B B_k)$$

has $G(B)$ as a bilimit with universal pseudo cone $\pi' : \Delta_{G(B)} \Rightarrow F$ as defined above.

One common way to define a stack is via descent objects as in [**17**], [**18**], [**42**], or [**52**].

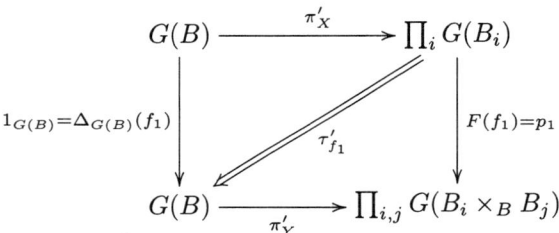

DEFINITION 12.4. Let $\mathcal{B}$ be a Grothendieck topology and $G : \mathcal{B} \to Cat$ a contravariant pseudo functor. Suppose that $\{B_i \to B\}_i$ is a Grothendieck cover. Then an *object with descent data on* $\{B_i \to B\}_i$ consists of an object $\{a_i\}_i \in \prod_{i \in I} G(B_i)$ and isomorphisms $\phi_{ij} : G(\pi^2_{ij})a_j \to G(\pi^1_{ij})a_i$ in $G(B_i \times_B B_j)$ which satisfy the *cocycle condition*

$$G(\pi^{13}_{ijk})\phi_{ik} = G(\pi^{12}_{ijk})\phi_{ij} \circ G(\pi^{23}_{ijk})\phi_{jk}$$

in $G(B_i \times_B B_j \times_B B_k)$ up to the coherence isos of the pseudo functor $G$. See below. A *morphism of descent objects* $\{\xi_i\}_i : \{a_i\}_i \to \{a'_i\}_i$ is a morphism in $\prod_{i \in I} G(B_i)$ such that the diagram

$$\begin{array}{ccc} G(\pi^2_{ij})a_j & \xrightarrow{\phi_{ij}} & G(\pi^1_{ij})a_i \\ \downarrow{\scriptstyle G(\pi^2_{ij})\xi_j} & & \downarrow{\scriptstyle G(\pi^1_{ij})\xi_i} \\ G(\pi^2_{ij})a'_j & \xrightarrow[\phi'_{ij}]{} & G(\pi^1_{ij})a'_i \end{array}$$

commutes in $G(B_i \times_B B_j)$. These objects and morphisms form the *category of descent data on the cover* $\{B_i \to B\}_i$ . This category is denoted $G(\{B_i \to B\}_i)$. There is a functor $G(B) \to G(\{B_i \to B\}_i)$ defined by $a \mapsto \{G(g_i)a\}_i$ where $g_i : B_i \to B$ are the morphisms from the Grothendieck cover. The $\phi_{ij}$ belonging to the image of $a$ under this functor are $\phi_{ij} := (\gamma^G_{\pi^1_{ij}, g_i} a)^{-1} \circ (\gamma^G_{\pi^2_{ij}, g_j} a)$.

The *cocycle condition* can be stated explicitly as the requirement that the following diagram commutes.

$$
\begin{array}{ccccc}
G(\pi^{23}_{ijk})G(\pi^{2}_{jk})a_k & \xrightarrow{G(\pi^{23}_{ijk})\phi_{jk}} & G(\pi^{23}_{ijk})G(\pi^{1}_{jk})a_j & \xrightarrow{\gamma_{\pi^{23}_{ijk},\pi^{1}_{jk}}a_j} & G(\pi^{1}_{jk}\circ\pi^{23}_{ijk})a_j \\
{\scriptstyle \gamma_{\pi^{23}_{ijk},\pi^{2}_{jk}}a_k}\downarrow & & & & \Vert \\
G(\pi^{2}_{jk}\circ\pi^{23}_{ijk})a_k & & & & G(\pi^{2}_{ij}\circ\pi^{12}_{ijk})a_j \\
\Vert & & & & \downarrow{\scriptstyle \gamma^{-1}_{\pi^{12}_{ijk},\pi^{2}_{ij}}a_j} \\
G(\pi^{2}_{ik}\circ\pi^{13}_{ijk})a_k & & & & G(\pi^{12}_{ijk})\circ G(\pi^{2}_{ij})a_j \\
{\scriptstyle \gamma^{-1}_{\pi^{13}_{ijk},\pi^{2}_{ik}}a_k}\downarrow & & & & \downarrow{\scriptstyle G(\pi^{12}_{ijk})\phi_{ij}} \\
G(\pi^{13}_{ijk})\circ G(\pi^{2}_{ik})a_k & & & & G(\pi^{12}_{ijk})\circ G(\pi^{1}_{ij})a_i \\
{\scriptstyle G(\pi^{13}_{ijk})\phi_{ik}}\downarrow & & & & \downarrow{\scriptstyle \gamma_{\pi^{12}_{ijk},\pi^{1}_{ij}}a_i} \\
G(\pi^{13}_{ijk})\circ G(\pi^{1}_{ik})a_i & \xleftarrow[\gamma^{-1}_{\pi^{13}_{ijk},\pi^{1}_{ik}}a_i]{} & G(\pi^{1}_{ik}\circ\pi^{13}_{ijk})a_i & = & G(\pi^{1}_{ij}\circ\pi^{12}_{ijk})a_i
\end{array}
$$

This diagram is another reason why we require our pseudo functors to have coherence arrows that are iso: if $\gamma$ were not invertible, the cocycle condition cannot be stated.

DEFINITION 12.5. If $\mathcal{B}$ is a Grothendieck topology, then a *Giraud stack of categories on* $\mathcal{B}$ is a contravariant pseudo functor $G : \mathcal{B} \to Cat$ such that for any object $B$ of $\mathcal{B}$ and any Grothendieck cover $\{B_i \to B\}_i$ of $B$, the functor $G(B) \to G(\{B_i \to B\}_i)$ is an equivalence of categories.[1]

THEOREM 12.6. *Let $G : \mathcal{B} \to Cat$ be a contravariant pseudo functor from a Grothendieck topology to the 2-category of small categories. Then $G$ is a stack if and only if it is a Giraud stack.*

*Proof:* From Chapter 5 we know that the category $L := PseudoCone(\mathbf{1}, F)$ is a pseudo limit of $F$. It is described as a subcategory of an appropriate product in Remarks 5.4 and 5.5 in such a way that the pseudo cone $\pi : \Delta_L \Rightarrow F$ consists of projections as in Remark 5.6.

We claim that the category $L$ of pseudo cones on a point is equivalent to the category $G(\{B_i \to B\}_i)$ of descent data by a functor $H : L \to G(\{B_i \to B\}_i)$.

[1]This is not standard terminology. We have only introduced it to distinguish the two definitions in the proof of their equivalence.

Recall from Remark 5.4 that each object of $L$ corresponds to a tuple

$$\{a_i\}_i \times \{a_{ij}\}_{ij} \times \{a_{ijk}\}_{ijk} \times \{\varepsilon_f\}_f$$

of objects

$$\{a_i\}_i \in \prod_i G(B_i),$$

$$\{a_{ij}\}_{ij} \in \prod_{ij} G(B_i \times_B B_j),$$

$$\{a_{ijk}\}_{ijk} \in \prod_{ijk} G(B_i \times_B B_j \times_B B_k),$$

and morphisms $\varepsilon_f$ indexed by morphisms $f$ of $\mathcal{J}$ appropriately. For example, $\varepsilon_{f_1} : F(f_1)\{a_i\}_i \to \{a_{ij}\}_{ij}$. These morphisms satisfy the two axioms listed in Remark 5.4. Each morphism in $L$ corresponds to a tuple

$$\{\xi_i\}_i \times \{\xi_{ij}\}_{ij} \times \{\xi_{ijk}\}_{ijk}$$

of morphisms in the product categories above and this tuple commutes with the morphisms $\varepsilon_f$ appropriately. Define

$$H(\{a_i\}_i \times \{a_{ij}\}_{ij} \times \{a_{ijk}\}_{ijk} \times \{\varepsilon_f\}_f) := \{a_i\}_i$$

$$H(\{\xi_i\}_i \times \{\xi_{ij}\}_{ij} \times \{\xi_{ijk}\}_{ijk}) := \{\xi_i\}_i.$$

The descent data for $\{a_i\}_i$ are defined as the components of $\{\phi_{ij}\}_{ij} := (\varepsilon_{f_1})^{-1} \circ \varepsilon_{f_2}$. Morphisms of $L$ map to morphisms of $G(\{B_i \to B\}_i)$ because the outer diagram of

(12.4)
$$\begin{array}{ccccc}
 & & \xleftarrow{\quad \{\phi_{ij}\}_{ij} \quad} & & \\
F(f_1)\{a_i\}_i & \xrightarrow{\varepsilon_{f_1}} & \{a_{ij}\}_{ij} & \xleftarrow{\varepsilon_{f_2}} & F(f_2)\{a_i\}_i \\
\downarrow{\scriptstyle F(f_1)\{\xi_i\}_i} & & \downarrow{\scriptstyle \{\xi_{ij}\}_{ij}} & & \downarrow{\scriptstyle F(f_2)\{\xi_i\}_i} \\
F(f_1)\{a'_i\}_i & \xrightarrow{\varepsilon'_{f_1}} & \{a'_{ij}\}_{ij} & \xleftarrow{\varepsilon'_{f_2}} & F(f_2)\{a'_i\}_i \\
 & & \xleftarrow{\quad \{\phi'_{ij}\}_{ij} \quad} & &
\end{array}$$

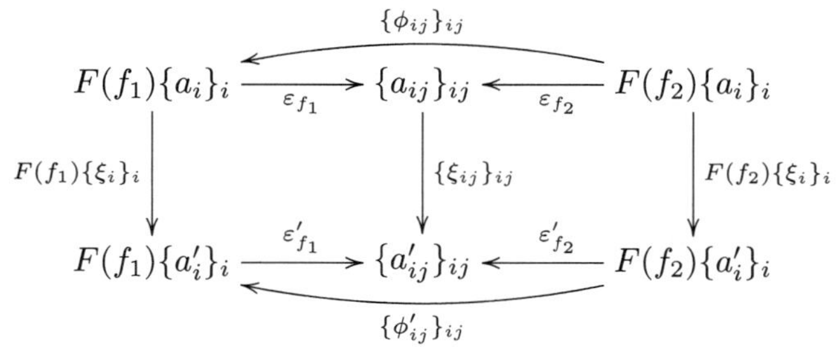

commutes by Remark 5.5. To see that the $\phi_{ij}$ satisfy the cocycle condition, consider the diagram below.

(12.5)

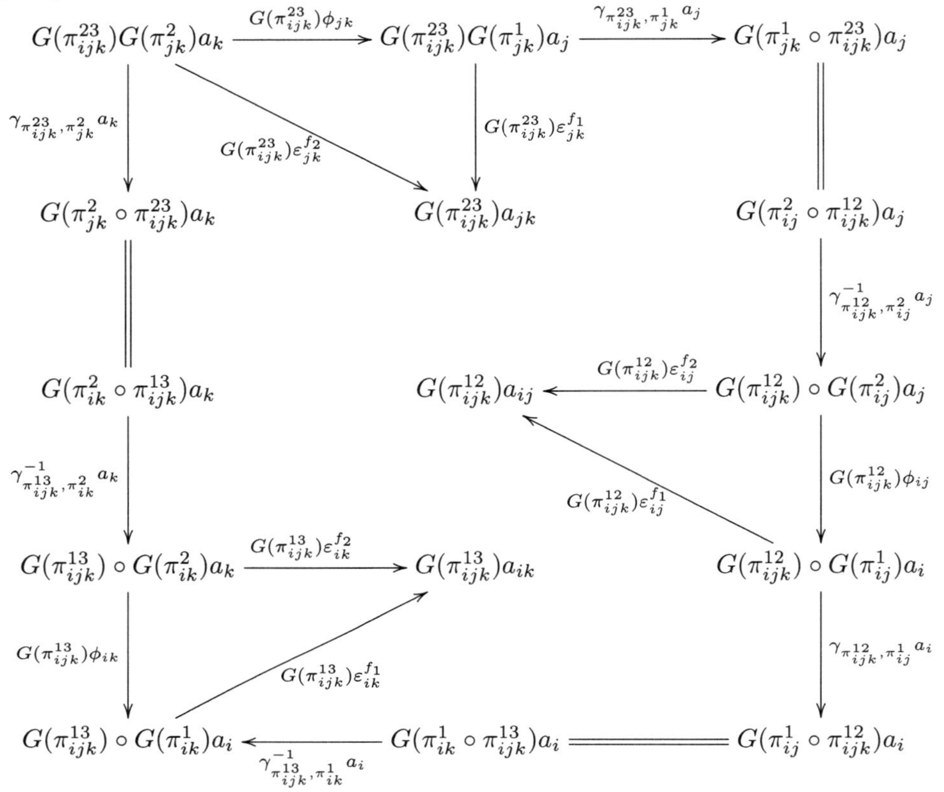


We want to show that the outer rectangle commutes. The small triangles commute by definition of $\phi_{ij}$. Next we draw another vertex $a_{ijk}$ inside the rectangle but outside the triangles. Then we draw the arrows $\varepsilon^f_{ijk}$ for all non-identity morphisms $F$ of the category $\mathcal{J}$ with target $Z$. All of these arrows terminate at $a_{ijk}$. Each of the resulting subdiagrams commutes because of the relations in $\mathcal{J}$ or because of the second axiom on the morphisms $\varepsilon_f$ in Remark 5.4 . Note that we are using the notation $\varepsilon_f = \{\varepsilon^f_{ijk}\}_{ijk}$. The outer rectangle commute because all of the subdiagrams commute and everything is iso. Hence the $\phi_{ij}$'s satisfy the cocycle condition and $H$ maps $L$ into $G(\{B_i \to B\}_i)$. These assignments obviously define a functor $H$.

The functor $H$ is faithful. Suppose

$$H(\{\xi_i\}_i \times \{\xi_{ij}\}_{ij} \times \{\xi_{ijk}\}_{ijk}) = H(\{\xi'_i\}_i \times \{\xi'_{ij}\}_{ij} \times \{\xi'_{ijk}\}_{ijk}).$$

Then $\{\xi_i\}_i = \{\xi'_i\}_i$. From this we conclude $\{\xi_{ij}\}_{ij} = \{\xi'_{ij}\}_{ij}$ by diagram (12.4). A similar diagram with objects $\{a_{ijk}\}$ and $\{a'_{ijk}\}$ in the center and arrows $\varepsilon_{f_{12}}, \varepsilon_{f_{23}}$ and $\varepsilon'_{f_{12}}, \varepsilon'_{f_{23}}$ pointing inward shows that $\{\xi_{ijk}\}_{ijk} = \{\xi'_{ijk}\}_{ijk}$.

The functor $H$ is also full. Let $\{\xi_i\}_i$ be a morphism in the category of descent data. Suppose further that its source and target lie in the image of $H$. Then the outer diagram of diagram (12.4) commutes and we define $\{\xi_{ij}\}_{ij}$ to be the unique arrow that makes diagram (12.4) commute. It exists because the horizontal arrows are iso. We can also define $\{\xi_{ijk}\}_{ijk}$ similarly, although we need to use diagram (12.4) several times and the naturality of $\gamma^G$ to show that the necessary diagrams in Remark 5.5 commute.

The functor $H$ is also surjective on objects. Suppose $\{a_i\}$ is an object with descent data $\phi_{ij}$. Define $a_{ij} := G(\pi^1_{ij})a_i$ and $a_{ijk} := G(\pi^1_{ik} \circ \pi^{13}_{ijk})a_i$. Define $\varepsilon^{f_1}_{ij}$ : $G(\pi^1_{ij})a_i \to a_{ij}$ to be the identity and $\varepsilon^{f_2}_{ij} := \phi_{ij}$. Let $\varepsilon^{f_{13}\circ f_1}_{ijk} : G(\pi^1_{ik} \circ \pi^{13}_{ijk})a_i \to a_{ijk}$ also be the identity. Any $\varepsilon$ indexed by an identity morphism is also trivial. Consider diagram (12.5) with the additional vertex $a_{ijk}$ and the additional $\varepsilon$'s mentioned just after diagram (12.5). Requiring the inner diagrams to commute uniquely defines the other $\varepsilon$'s which we did not define yet. The commutivity of these smaller diagrams guarantees that the tuple

$$\{a_i\}_i \times \{a_{ij}\}_{ij} \times \{a_{ijk}\}_{ijk} \times \{\varepsilon_f\}_f$$

we have just defined is an object of $L$. This object obviously maps under $H$ to $\{a_i\}_i$ with the correct descent data.

We conclude $H$ is an equivalence because it is faithfully full and essentially surjective. Hence the category $L$ of pseudo cones is equivalent to the category $G(\{B_i \to B\}_i)$ of descent data.

There is also a functor $G(B) \to L$ defined like the functor $G(B) \to G(\{B_i \to B\}_i)$ that makes the diagrams

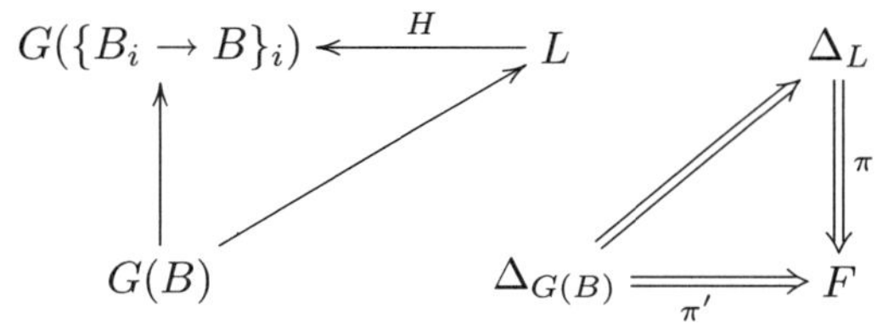


commute. Suppose $G$ is a Giraud stack. Then the left vertical arrow is an equivalence. Hence the functor $G(B) \to L$ is an equivalence and $\pi'$ makes $G(B)$ into a bilimit of $F$ because $L$ is a bilimit of $F$ with pseudo limiting cone $\pi$. Hence $G$ is a stack.

Suppose $G$ is a stack. Then $\pi'$ makes $G(B)$ into a bilimit of $F$. Then the functor $G(B) \to L$ is an equivalence because $L$ is also a bilimit and the right diagram commutes. Hence the functor $G(B) \to G(\{B_i \to B\}_i)$ is also an equivalence and $G$ is a Giraud stack.

This completes the proof that the two definitions of stack are equivalent.

□

Lastly, we define stacks of objects in a 2-category which admits bilimits, such as the 2-category of pseudo algebras over a theory.

DEFINITION 12.7. Let $\mathcal{C}$ be a 2-category whose objects have underlying categories. Suppose $\mathcal{B}$ is a Grothendieck topology and $\mathcal{C}$ admits bilimits. A *stack of objects of* $\mathcal{C}$ is a contravariant pseudo functor $G : \mathcal{B} \to \mathcal{C}$ which takes Grothendieck covers to bilimits, *i.e.* for any object $B$ of $\mathcal{B}$ and any Grothendieck cover $\{g_i : B_i \to B | i \in I\} \in K(B)$ the diagram

$$\prod_{i\in I} G(B_i) \underset{p_2}{\overset{p_1}{\rightrightarrows}} \prod_{i,j\in I} G(B_i \times_B B_j) \overset{p_{12}}{\underset{p_{23}}{\overset{p_{13}}{\Rrightarrow}}} \prod_{i,j,k\in I} G(B_i \times_B B_j \times_B B_k)$$

has $G(B)$ as a bilimit with universal pseudo cone $\pi' : \Delta_{G(B)} \Rightarrow F$ as defined above.

For example, a stack of pseudo algebras over a theory $T$ is a contravariant pseudo functor from a Grothendieck topology into the 2-category of pseudo $T$-algebras which takes Grothendieck covers to bilimits in the above sense.

CHAPTER 13

# 2-Theories, Algebras, and Weighted Pseudo Limits

The algebraic structure of the category of rigged surfaces can be described as a pseudo algebra over a certain 2-theory as in [**25**], [**26**], and [**27**]. A *pseudo algebra over a 2-theory* in this paper is the same as a *lax algebra over a 2-theory* in [**25**], [**26**], and [**27**]. However, the 2-theories of [**53**], [**54**], and [**55**] are different from the 2-theories in this paper. In this chapter we review the relevant terminology and prove results about limits. Before giving the definition of a 2-theory, we motivate it with an example in the first section.

## 13.1. The 2-Theory *End*($X$) Fibered over the Theory *End*($I$)

Let $I$ be a category and $k$ a positive integer. Suppose $X : I^k \longrightarrow Cat$ is a strict 2-functor from the category $I^k$ to the 2-category $Cat$ of small categories. Here $I^k$ is interpreted as a 2-category where the hom sets are discrete categories. We will now describe the *2-theory End(X) fibered over the theory End(I)*, which is a contravariant functor $End(I) \longrightarrow Cat$ satisfying certain properties.

Recall that the theory $End(I)$ is the category with objects $0 = \{*\}, 1 = I, 2 = I^2, 3 = \dots$ and morphisms $Mor_{End(I)}(m,n) = Functors(I^m, I^n)$. Here $\{*\}$ denotes the terminal object in the category of small categories. As with any theory, the theory $End(I)$ can be completely described by the sets $End(I)(n) := Mor_{End(I)}(n,1)$, a composition, substitution, and a unit which satisfy a list of axioms. See Theorem 6.10 or [**25**] for details.

From the theory $End(I)$ we can obtain another category denoted $End(I)^k$, which also turns out to be a theory. It has objects $0 = \{*\} \times \cdots \times \{*\}, 1 = I \times \cdots \times I, 2 = I^2 \times \cdots \times I^2, 3 = \dots$ ($k$ copies in each product) and it has morphisms $Mor_{End(I)^k}(m,n) := Mor_{End(I)}(m,n)^{\times k}$. For example, $v \in Mor_{End(I)^k}(m,1)$ is a functor $v : (I^m)^k \longrightarrow I^k$ that is a $k$-tuple of functors $I^m \longrightarrow I$. For $n \in \mathbb{N}$ and $1 \leq i \leq n$, let $pr_i^{\times k} : (I^n)^k \longrightarrow I^k$ be the morphism $pr_i^{\times k} \in Mor_{End(I)^k}(n,1)$ whose $k$ components are each the projection functor $pr_i : I^n \longrightarrow I$ onto the $i$-th coordinate. We can easily check that $n \in Obj\ End(I)^k$ is the product in $End(I)^k$ of $n$ copies of 1 with projection morphisms $pr_1^{\times k}, \dots, pr_n^{\times k}$. Hence $End(I)^k$ is itself a theory and $Mor_{End(I)^k}(m,n)$ is in bijective correspondence with $\prod_{i=1}^n Mor_{End(I)^k}(m,1)$. We identify these two sets via the usual bijection. In other words, for $k$-tuples $w_1, \dots, w_n \in Mor_{End(I)^k}(m,1)$ we let $\prod_{j=1}^n w_j$ denote the unique morphism $m \longrightarrow n$ of $End(I)^k$ such that

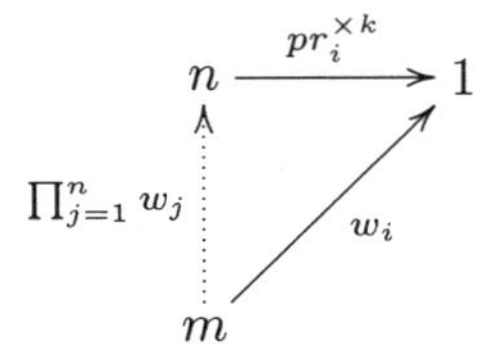

commutes for all $i = 1, \ldots, n$. This notation differs from [**25**], in which the notation $(w_1, \ldots, w_n)$ is used instead of the product. We reserve $(w_1, \ldots, w_n)$ for a different morphism. The reason for our choice will become clear later. Using our convention, we have $w = \prod_{j=1}^{n} pr_j^{\times k} \circ w$ for $w \in Mor_{End(I)^k}(m, n)$.

Since $End(I)^k$ is a theory, it has a substitution and a composition with unit which satisfy certain axioms described in Chapter 6 and [**25**]. If $f : \{1, \ldots, p\} \to \{1, \ldots, q\}$ is a function and $w \in End(I)^k(p) = Mor_{End(I)^k}(p, 1) = Mor_{End(I)}(p, 1)^{\times k}$, then the substituted word $w_f$ is obtained by substituting by $f$ in each of the words in the $k$-components of $w$. The composition is also done componentwise. The unit $1^{\times k} : I \times \cdots \times I \to I \times \cdots \times I$ is $k$ copies of the unit $1 : I \to I$ in the theory $End(I)$. These explicit descriptions of substitution, composition, and unit follow from the definitions of the projections in the theory $End(I)^k$ by the work in Chapter 6.

We follow the conventions of Chapter 6 to define a morphism $(w_1, \ldots, w_n)$. Let $w_i \in End(I)^k(m_i)$ for $i = 1, \ldots, n$. Let $\iota_i : \{1, \ldots, m_i\} \to \{1, \ldots, m_1 + m_2 + \cdots + m_n\}$ be the injective function which takes the domain to the $i$-th block. Then there exists a unique morphism $(w_1, \ldots, w_n)$ such that

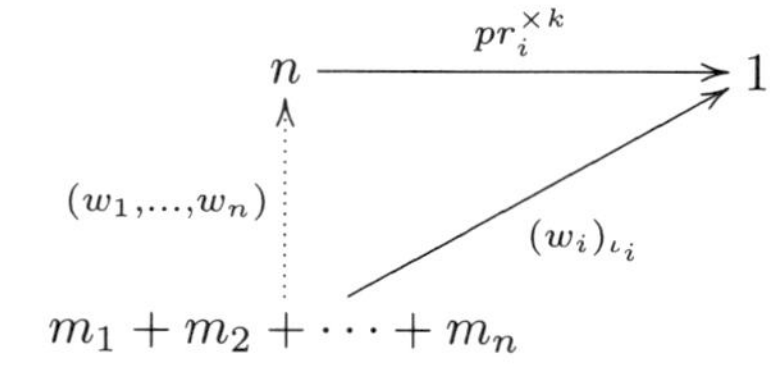


commutes for all $i = 1, \ldots, n$. Explicitly, the morphism $(w_1, \ldots, w_n)$ is obtained by doing an analogous process in each of the $k$ components.

The strict 2-functor $X : I^k \to Cat$ gives rise to a contravariant functor $End(X) : End(I) \to Cat$ as follows. For $m \in Obj\ End(I)$ the category $End(X)(m)$ has objects $Obj\ End(X)(m) = \coprod_{n \geq 0} Mor_{End(I)^k}(m, n)$, in other words, the objects of $End(X)(m)$ are the arrows of $End(I)^k$ with domain $m$. For $\prod_{i=1}^{p} v_i, \prod_{i=1}^{q} w_i \in Obj\ End(X)(m)$ where $v_1, \ldots, v_p, w_1, \ldots, w_q \in Mor_{End(I)^k}(m, 1)$ we define the set of morphisms $Mor_{End(X)(m)}(\prod_{i=1}^{p} v_i, \prod_{i=1}^{q} w_i)$ to be the collection of natural transformations

$$\alpha : X \circ v_1 \circ d^m \times \cdots \times X \circ v_p \circ d^m \Rightarrow X \circ w_1 \circ d^m \times \cdots \times X \circ w_q \circ d^m \tag{13.1}$$

where $d^m : I^m \to (I^m)^k$ is the diagonal functor. Note that $X \circ v_1 \circ d^m \times \cdots \times X \circ v_p \circ d^m$ and $X \circ w_1 \circ d^m \times \cdots \times X \circ w_q \circ d^m$ are functors $I^m \to Cat$. The composition of morphisms in $End(X)(m)$ is the vertical composition of natural transformations. With these definitions, $End(X)(m)$ is a category. We must still define the contravariant functor $End(X)$ on morphisms and verify that it preserves identities and compositions. For any morphism $u : I^\ell \to I^m$ of the theory $End(I)$, define $u^{\times k} : (I^\ell)^k \to (I^m)^k$ to be the functor which is $u$ in each of the $k$ components. Note that $u^{\times k} \circ d^\ell = d^m \circ u : I^\ell \to (I^m)^k$. The functor $End(X)(u) : End(X)(m) \to End(X)(\ell)$ is defined on objects by $End(X)(u)(\prod_{i=1}^{p} v_i) := \prod_{i=1}^{p} v_i \circ u^{\times k}$ and on morphisms $\alpha$ in (13.1) by $End(X)(u)(\alpha) := \alpha * i_u$ where $*$ denotes the horizontal composition of natural transformations and $i_u : u \Rightarrow u$ is the trivial natural transformation. This makes sense because

$$\begin{aligned}(X \circ v_1 \circ d^m \times \cdots \times X \circ v_p \circ d^m) \circ u &= X \circ v_1 \circ d^m \circ u \times \cdots \times X \circ v_p \circ d^m \circ u \\ &= X \circ v_1 \circ u^{\times k} \circ d^\ell \times \cdots \times X \circ v_p \circ u^{\times k} \circ d^\ell\end{aligned}$$

and

$$\alpha * i_u : X \circ v_1 \circ u^{\times k} \circ d^\ell \times \cdots \times X \circ v_p \circ u^{\times k} \circ d^\ell \Rightarrow X \circ w_1 \circ u^{\times k} \circ d^\ell \times \cdots \times X \circ w_q \circ u^{\times k} \circ d^\ell$$

really is a morphism

$$End(X)(u)(\prod_{i=1}^{p} v_i) = \prod_{i=1}^{p} v_i \circ u^{\times k} \to \prod_{i=1}^{p} w_i \circ u^{\times k} = End(X)(u)(\prod_{i=1}^{q} w_i).$$

If $u : I^\ell \to I^m$ is the identity functor $I^m \to I^m$, then $End(X)(u) : End(X)(m) \to End(X)(m)$ is also the identity functor because $v_i \circ u^{\times k} = v_i$ for $i = 1, \ldots, p$ and $w_i \circ u^{\times k} = w_i$ for $i = 1, \ldots, q$ and also $\alpha * i_u = \alpha$. If $I^j \xrightarrow{u_1} I^m \xrightarrow{u_2} I^\ell$ are morphisms in *End*($I$), then $u_2^{\times k} \circ u_1^{\times k} = (u_2 \circ u_1)^{\times k}$ and

$$(\alpha * i_{u_2}) * i_{u_1} = \alpha * (i_{u_2} * i_{u_1}) = \alpha * i_{u_2 \circ u_1},$$

which together imply that

$$End(X)(u_2 \circ u_1) = End(X)(u_1) \circ End(X)(u_2).$$

Thus $End(X) : End(I) \to Cat$ preserves identities and compositions and is a contravariant functor.

The category $End(X)(m)$ also admits certain products, which will be a feature of a general 2-theory. For $v_1, \ldots, v_p \in Mor_{End(I)^k}(m, 1)$ and $\prod_{i=1}^p v_i \in Mor_{End(I)^k}(m, p) \subseteq Obj\ End(X)(m)$ define projections $pr_j : \prod_{i=1}^p v_i \to v_j$ for $j = 1, \ldots, p$ to be the projection natural transformations

$$X \circ v_1 \circ d^m \times \cdots \times X \circ v_p \circ d^m \Rightarrow X \circ v_j \circ d^m.$$

Then $\prod_{i=1}^p v_i$ is obviously the product of $v_1, \ldots, v_p$ in the category $End(X)(m)$ with these projections. This explains the choice of notation $\prod_{i=1}^p v_i$. This product property will also be required of a general 2-theory. We record for later use how these products allow us to define morphisms $\iota'$ for every function $\iota : \{1, \ldots, p\} \to \{1, \ldots, q\}$. Let $w_1, \ldots, w_q \in Mor_{End(I)^k}(m, 1) \subseteq Obj\ End(X)(m)$. Then for a function $\iota : \{1, \ldots, p\} \to \{1, \ldots, q\}$ there exists a unique morphism $\iota'$ such that

(13.2)
$$\begin{array}{ccc} \prod_{i=1}^p w_{\iota(i)} & \xrightarrow{pr_\ell} & w_\ell \\ \uparrow{\scriptstyle \iota'} & \nearrow{\scriptstyle pr_{\iota(\ell)}} & \\ \prod_{i=1}^q w_i & & \end{array}$$

commutes for all $\ell = 1, \ldots, p$. The arrows of the natural transformation $\iota' : X \circ w_1 \circ d^m \times \cdots \times X \circ w_q \circ d^m \Rightarrow X \circ w_{\iota(1)} \circ d^m \times \cdots \times X \circ w_{\iota(p)} \circ d^m$ have the appropriate projections as their components.

The 2-theory *End*($X$) has several operations on it which any general 2-theory will also have, once we define the notion of 2-theory. To make the description of these operations easier, we follow the notation introduced by P. Hu and I. Kriz in [**25**]. For objects $w, w_1, \ldots, w_q \in Mor_{End(I)^k}(m, 1) \subseteq Obj\ End(X)(m)$ we set

$$End(X)(w; w_1, \ldots, w_q) := Mor_{End(X)(m)}(\prod_{i=1}^{q} w_i, w).$$

The operations of P. Hu and I. Kriz are collated in the following theorem.

THEOREM 13.1. *The contravariant functor $End(X) : End(I) \to Cat$ has the following operations.*

(1) *For each $w \in T^k(m)$ there exists a unit $1_w \in End(X)(w; w)$.*
(2) *For all $w, w_i, w_{ij} \in Mor_{End(I)^k}(m, 1)$ there is a function called $End(X)$-composition.*

$$\gamma : End(X)(w; w_1, \ldots, w_q) \times End(X)(w_1; w_{11}, \ldots, w_{1p_1}) \times \cdots \times End(X)(w_q; w_{q1}, \ldots, w_{qp_q})$$

$$\to End(X)(w; w_{11}, \ldots, w_{qp_q})$$

(3) *Let $w, w_1, \ldots, w_q \in Mor_{End(I)^k}(m, 1)$. For any function $\iota : \{1, \ldots, p\} \to \{1, \ldots, q\}$ there is a function*

$$()^\iota : End(X)(w; w_{\iota(1)}, \ldots, w_{\iota(p)}) \to End(X)(w; w_1, \ldots, w_q)$$

*called $End(X)$-functoriality.*

(4) *Let $w, w_1, \ldots, w_q \in Mor_{End(I)^k}(m, 1)$. For any function $f : \{1, \ldots, m\} \to \{1, \ldots, \ell\}$ there is a function*

$$()_f : End(X)(w; w_1, \ldots, w_q) \to End(X)(w_f; (w_1)_f, \ldots, (w_q)_f)$$

*where $w_f$ means to substitute $f$ in each of the words in the $k$-tuple $w$. This function is called $End(I)$-functoriality. Note that $End(X)(w; w_1, \ldots, w_q)$ is a hom set in the category $End(X)(m)$ while on the other hand $End(X)(w_f; (w_1)_f, \ldots, (w_q)_f)$ is a hom set in the category $End(X)(\ell)$.*

(5) *For $u_i \in End(I)(k_i), i = 1, \ldots, m$ and $w, w_1, \ldots, w_q \in Mor_{End(I)^k}(m, 1)$ let $v_j := \gamma^{\times k}(w_j; u_1^{\times k}, \ldots, u_m^{\times k})$ for $j = 1, \ldots, q$ and furthermore let $v := \gamma^{\times k}(w; u_1^{\times k}, \ldots, u_m^{\times k})$. Then there is a function*

$$(u_1, \ldots, u_m)^* : End(X)(w; w_1, \ldots, w_q) \to End(X)(v; v_1, \ldots, v_q)$$

*called $End(I)$-substitution. Here $\gamma^{\times k}$ means to use the composition of the theory $End(I)$ in each of the $k$ components, which coincides with composition in the theory $End(I)^k$. Note that $End(X)(w; w_1, \ldots, w_q)$ is a hom set in the category $End(X)(m)$ while $End(X)(v; v_1, \ldots, v_q)$ is a hom set in the category $End(X)(k_1 + \cdots + k_m)$.*

*Proof:*

(1) The unit $1_w : X \circ w \circ d^m \Rightarrow X \circ w \circ d^m$ is the identity natural transformation $i_{X \circ w \circ d^m} : X \circ w \circ d^m \Rightarrow X \circ w \circ d^m$.
(2) Let $\alpha : \prod_{i=1}^q w_i \to w$ and $\alpha_i : \prod_{j=1}^{p_i} w_{ij} \to w_i$ for $i = 1, \ldots, q$ be morphisms of $End(X)(m)$. Let $\iota_\ell : \{1, \ldots, p_\ell\} \to \{1, \ldots, p_1 + p_2 + \cdots + p_q\}$ be the injective function which takes the domain to the $\ell$-th block. We take the product $\prod_{i=1}^q \prod_{j=1}^{p_i} w_{ij}$ to be

$$\prod_{i=1}^q \prod_{j=1}^{p_i} w_{ij} = w_{11} \times w_{12} \times \cdots w_{1p_1} \times w_{21} \times \cdots \times w_{2p_2} \times w_{31} \times \cdots \times w_{qp_q}.$$

Then there exists a unique morphism $(\alpha_1, \ldots, \alpha_q)$ such that

$$\begin{array}{ccc} \prod_{i=1}^q w_i & \xrightarrow{pr_\ell^{\times k}} & w_\ell \\ {\scriptstyle (\alpha_1,\ldots,\alpha_q)} \uparrow & \nearrow {\scriptstyle (\alpha_\ell)_{\iota_\ell}} & \\ \prod_{i=1}^q \prod_{j=1}^{p_i} w_{ij} & & \end{array}$$

commutes for all $\ell = 1, \ldots, q$. This means that

$$(\alpha_1, \ldots, \alpha_q) : X \circ w_{11} \circ d^m \times \cdots \times X \circ w_{qp_q} \circ d^m \Rightarrow$$
$$X \circ w_1 \circ d^m \times X \circ w_2 \circ d^m \times \cdots \times X \circ w_q \circ d^m$$

is the natural transformation which is $\alpha_\ell$ on $X \circ w_{\ell 1} \circ d^m \times \cdots \times X \circ w_{\ell p_\ell} \circ d^m$. Define

$$\gamma(\alpha; \alpha_1, \ldots, \alpha_q) := \alpha \circ (\alpha_1, \ldots \alpha_q)$$

where the composition is in the category $End(X)(m)$.

(3) Let $w_1, \ldots, w_q \in Mor_{End(I)^k}(m, 1)$ and $\iota : \{1, \ldots, p\} \to \{1, \ldots, q\}$ be a function. Let $\iota' : \prod_{i=1}^q w_i \to \prod_{i=1}^q w_{\iota(i)}$ be the morphism defined in diagram (13.2). Then we define $End(X)$-functoriality

$$End(X)(w; w_{\iota(1)}, \ldots, w_{\iota(p)}) \to End(X)(w; w_1, \ldots, w_q)$$

by $\alpha \mapsto \alpha \circ \iota'$.

(4) A function $f : \{1, \ldots, m\} \to \{1, \ldots, \ell\}$ induces a morphism $f' : \ell \to m$ in $End(I)$ which in turn gives rise to a morphism $(f')^{\times k} : (I^\ell)^k \to (I^m)^k$ in $End(I)^k$. Then $w_f = w \circ (f')^{\times k}$ by definition and the functor $End(X)(f') : End(X)(m) \to End(X)(\ell)$ gives us a map of hom sets

$$()_f : End(X)(w; w_1, \ldots, w_q) \to End(X)(w_f; (w_1)_f, \ldots, (w_q)_f).$$

(5) Let $\iota_i : \{1, \ldots, k_i\} \to \{1, \ldots, k_1 + k_2 + \cdots + k_m\}$ be the injective map which takes the domain to the $i$-th block. Let $(u_1^{\times k}, \ldots, u_m^{\times k})$ denote the unique morphism in $End(I)^k$ such that

$$\begin{array}{ccc} m & \xrightarrow{pr_i^{\times k}} & 1 \\ {\scriptstyle (u_1^{\times k},\ldots,u_m^{\times k})} \uparrow & \nearrow {\scriptstyle (u_i)_{\iota_i}^{\times k}} & \\ k_1 + k_2 + \cdots + k_m & & \end{array}$$

commutes. Then we know from the general theory of theories that $\gamma^{\times k}(w; u_1^{\times k}, \ldots, u_m^{\times k}) = w \circ (u_1^{\times k}, \ldots, u_m^{\times k})$ where the composition "$\circ$" is the composition in the category $End(I)^k$. Then $End(X)(u_1^{\times k}, \ldots, u_m^{\times k})(w) = v$ and the functor $End(X)(u_1^{\times k}, \ldots, u_m^{\times k})$ gives us the desired map of hom sets.

□

These operations on $End(X)$ satisfy certain relations.

THEOREM 13.2. *The operations on the contravariant functor $End(X) : End(I) \to Cat$ satisfy the following relations.*

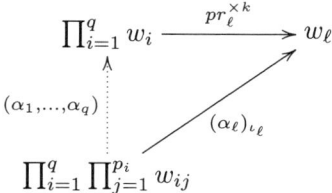

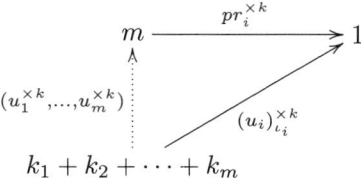

(1) *End(X)-composition is associative, i.e.* $\gamma(\alpha;\gamma(\alpha^1;\alpha^1_1,\dots,\alpha^1_{n_1}),\gamma(\alpha^2;\alpha^2_1,\dots,\alpha^2_{n_2}),\dots,\gamma(\alpha^q;\alpha^q_1,\dots,\alpha^q_{n_q}))$ *is the same as* $\gamma(\gamma(\alpha;\alpha^1,\dots,\alpha^q),\alpha^1_1,\dots\alpha^1_{n_1},\alpha^2_1,\dots,\alpha^2_{n_2},\dots,\alpha^q_1,\dots,\alpha^q_{n_q})$.
(2) *End(X)-composition is unital, i.e. for* $\alpha \in End(X)(w;w_1,\dots,w_q)$ *we have* $\gamma(\alpha;1_{w_1},\dots,1_{w_q}) = \alpha = \gamma(1_w;\alpha)$.
(3) *End(X)-functoriality is functorial, i.e. for functions* $\{1,\dots,p\}\xrightarrow{\iota}\{1,\dots,q\}\xrightarrow{\theta}\{1,\dots,r\}$ *the composition*

$$End(X)(w;w_{\theta\iota(1)},\dots,w_{\theta\iota(p)}) \xrightarrow{()^\iota} End(X)(w;w_{\theta(1)},\dots,w_{\theta(q)})$$

$$\xrightarrow{()^\theta} End(X)(w;w_1,\dots,w_r)$$

*is the same as*

$$End(X)(w;w_{\theta\iota(1)},\dots,w_{\theta\iota(p)})\xrightarrow{()^{\theta\circ\iota}}End(X)(w_1,\dots,w_r)$$

*and for the identity* $id_q:\{1,\dots,q\}\to\{1,\dots,q\}$ *the map* $()^{id_q}: End(X)(w;w_1,\dots,w_q)\to End(X)(w;w_1,\dots,w_q)$ *is the identity.*
(4) *The End(X)-compositions* $\gamma$ *are equivariant with respect to End(X)-functoriality in the sense that if* $\iota:\{1,\dots,p\}\to\{1,\dots,q\}$ *is a function,* $\alpha\in End(X)(w;w_{\iota(1)},\dots,w_{\iota(p)})$*, and* $\alpha_\ell \in End(X)(w_\ell;w_{\ell 1},\dots,w_{\ell p_\ell})$ *for* $\ell=1,\dots,q$ *then*

$$\gamma(\alpha^\iota;\alpha_1,\dots,\alpha_q) = \gamma(\alpha;\alpha_{\iota(1)},\dots,\alpha_{\iota(p)})^{\bar{\iota}},$$

*where* $\bar{\iota}:\{1,2,\dots,p_{\iota(1)}+\cdots+p_{\iota(p)}\}\to\{1,2,\dots,p_1+\cdots+p_q\}$ *is the function obtained by parsing the sequence* $1,2,\dots,p_1+\cdots+p_q$ *into consecutive blocks* $B_1,\dots,B_q$ *of lengths* $p_1,\dots,p_q$ *and then writing them in the order* $B_{\iota(1)},\dots,B_{\iota(p)}$ *as in Example 6.3.*
(5) *The End(X)-compositions* $\gamma$ *are equivariant with respect to End(X)-functoriality in the sense that if* $\alpha\in End(X)(w;w_1,\dots,w_q)$*,* $\alpha_\ell\in End(X)(w_\ell;w_{\ell\iota_\ell(1)},\dots,w_{\ell\iota_\ell(p'_\ell)})$*, and* $\iota_\ell:\{1,\dots,p'_\ell\}\to\{1,\dots,p_\ell\}$ *are functions for* $\ell=1,\dots,q$ *then*

$$\gamma(\alpha;(\alpha_1)^{\iota_1},\dots,(\alpha_q)^{\iota_q}) = \gamma(\alpha;\alpha_1,\dots,\alpha_q)^{\iota_1+\cdots\iota_q}$$

*where* $\iota_1+\cdots+\iota_q:\{1,\dots,p'_1+\cdots+p'_q\}\to\{1,\dots,p_1+\cdots+p_q\}$ *is the function obtained by placing* $\iota_1,\dots,\iota_q$ *side by side.*
(6) *End(I)-functoriality is functorial, i.e.*

*for functions* $\{1,\dots,n\}\xrightarrow{f}\{1,\dots,m\}\xrightarrow{g}\{1,\dots,\ell\}$ *and words* $w,w_1,\dots,w_q\in Mor_{End(I)^k}(n,1)$ *the composition*

$$End(X)(w;w_1,\dots,w_q)\xrightarrow{()_f}End(X)(w_f;(w_1)_f,\dots,(w_q)_f)$$

$$\xrightarrow{()_g}End(X)((w_f)_g;((w_1)_f)_g,\dots,((w_q)_f)_g)$$

*is the same as*

$$End(X)(w;w_1,\dots,w_q)\xrightarrow{()_{g\circ f}}End(X)(w_{g\circ f};(w_1)_{g\circ f},\dots,(w_q)_{g\circ f})$$

*and for the identity* $id_n:\{1,\dots n\}\to\{1,\dots,n\}$ *the map* $()_{id_n}: End(X)(w;w_1,\dots,w_q)\to End(X)(w;w_1,\dots,w_q)$ *is the identity.*

(7) *$End(I)$-substitution is associative.*
*Let $w, w_1, \ldots, w_q \in Mor_{End(I)^k}(m,1)$, $t_i \in End(I)(k_i)$ for $i = 1, \ldots, m$ and $s_{ij} \in End(I)(k_{ij})$ for $1 \leq i \leq m$ and $1 \leq j \leq k_i$. Let*

$$v := \gamma^{\times k}(w; t_1^{\times k}, \ldots, t_m^{\times k})$$
$$v_\ell := \gamma^{\times k}(w_\ell; t_1^{\times k}, \ldots, t_m^{\times k})$$
$$u := \gamma^{\times k}(v; s_{11}^{\times k}, s_{12}^{\times k}, \ldots, s_{1k_1}^{\times k}, s_{21}^{\times k}, \ldots, s_{31}^{\times k}, \ldots, s_{m1}^{\times k}, \ldots, s_{mk_m}^{\times k})$$
$$u_\ell := \gamma^{\times k}(v_\ell; s_{11}^{\times k}, s_{12}^{\times k}, \ldots, s_{1k_1}^{\times k}, s_{21}^{\times k}, \ldots, s_{31}^{\times k}, \ldots, s_{m1}^{\times k}, \ldots, s_{mk_m}^{\times k})$$

*for $\ell = 1, \ldots, q$. Then the composition*

$$End(X)(w; w_1, \ldots, w_q) \xrightarrow{(t_1,\ldots,t_m)^*} End(X)(v; v_1, \ldots, v_q)$$

$$\xrightarrow{(s_{11},\ldots,s_{mk_m})^*} End(X)(u; u_1, \ldots u_q)$$

*is the same as*

$$End(X)(w; w_1, \ldots, w_q) \xrightarrow{(r_1,\ldots,r_m)^*} End(X)(u; u_1, \ldots, u_q)$$

*where $r_i = \gamma^{\times k}(t_i^{\times k}; s_{i1}^{\times k}, s_{i2}^{\times k}, \ldots, s_{ik_i}^{\times k}) = \gamma_{End(I)}(t_i; s_{i1}, s_{i2}, \ldots, s_{ik_i})^{\times k}$ for $i = 1, \ldots, m$. Note that $u =$ $\gamma^{\times k}(w; \gamma^{\times k}(t_1^{\times k}; s_{11}^{\times k}, s_{12}^{\times k}, \ldots, s_{1k_1}^{\times k}), \ldots, \gamma^{\times k}(t_m^{\times k}; s_{m1}^{\times k}, s_{m2}^{\times k}, \ldots, s_{mk_m}^{\times k}))$.*

(8) *$End(I)$-substitution is unital.*
*For the unit $1 \in End(I)(1)$ of the theory $End(I)$ and $w, w_1, \ldots, w_q \in Mor_{End(I)^k}(m,1)$ the function*

$$(1, \ldots, 1)^* : End(X)(w; w_1, \ldots, w_q) \to End(X)(w; w_1, \ldots, w_q)$$

*is the identity.*

(9) *$End(X)$-composition is $End(I)$-equivariant.*
*If $f : \{1, \ldots, m\} \to \{1, \ldots, \ell\}$ is a function, $w, w_i, w_{ij} \in Mor_{End(I)^k}(m,1)$, $\alpha \in End(X)(w; w_1, \ldots w_q)$, and $\alpha_j \in End(X)(w_j; w_{j1}, \ldots, w_{jp_j})$ for $j = 1, \ldots, q$, then*

$$\gamma(\alpha_f; (\alpha_1)_f, \ldots, (\alpha_q)_f) = \gamma(\alpha; \alpha_1, \ldots, \alpha_q)_f.$$

(10) *$End(X)$-functoriality and $End(I)$-functoriality commute.*
*For functions $\iota : \{1, \ldots, p\} \to \{1, \ldots, q\}$ and $f : \{1, \ldots, m\} \to \{1, \ldots, \ell\}$ and morphism $\alpha \in End(X)(w; w_{\iota(1)}, \ldots, w_{\iota(p)})$ we have $(\alpha^\iota)_f = (\alpha_f)^\iota$.*

(11) *$End(X)$-functoriality and $End(I)$-substitution commute.*
*The diagram*

$$\begin{array}{ccc} End(X)(w; w_{\iota(1)}, \ldots, w_{\iota(p)}) & \xrightarrow{()^\iota} & End(X)(w; w_1, \ldots, w_q) \\ \downarrow{\scriptstyle (u_1,\ldots,u_m)^*} & & \downarrow{\scriptstyle (u_1,\ldots,u_m)^*} \\ End(X)(v; v_{\iota(1)}, \ldots, v_{\iota(p)}) & \xrightarrow[()^\iota]{} & End(X)(v; v_1, \ldots, v_q) \end{array}$$

*commutes.*

(12) *End(I)-functoriality and End(I)-substitution commute, in the sense that if* $f_i : \{1,\dots,k_i\} \to \{1,\dots,k_i'\}$ *are functions and* $u_i \in End(I)(k_i)$ *for* $i = 1,\dots,m$ *and* $w, w_1, \dots, w_q \in End(I)^k(m)$, *then the diagram below commutes.*

$$\begin{array}{ccc} End(X)(w; w_1,\dots,w_q) & \xrightarrow{(u_1,\dots,u_m)^*} & End(X)(v; v_1,\dots,v_q) \\ & \searrow{\scriptstyle ((u_1)_{f_1},\dots,(u_m)_{f_m})^*} & \downarrow{\scriptstyle ()_{f_1+\dots+f_m}} \\ & & End(X)(v_{f_1+\dots+f_m}; (v_1)_{f_1+\dots+f_m},\dots,(v_q)_{f_1+\dots+f_m}) \end{array}$$

*Note that*

$$\gamma^{\times k}(w; (u_1)^{\times k}_{f_1}, \dots, (u_m)^{\times k}_{f_m}) = \gamma^{\times k}(w; u_1,\dots,u_m)_{f_1+\dots+f_m}$$
$$= v_{f_1+\dots+f_m}.$$

(13) *End(I)-functoriality and End(I)-substitution commute, in the sense that if* $f : \{1,\dots,m\} \to \{1,\dots,\ell\}$ *is a function and* $u_i \in End(I)(k_i)$ *for* $i = 1,\dots,\ell$, *then the diagram*

$$\begin{array}{ccc} End(X)(w; w_1,\dots,w_q) & \xrightarrow{()_f} & End(X)(w_f; (w_1)_f,\dots,(w_q)_f) \\ \downarrow{\scriptstyle (u_{f1},\dots,u_{fm})^*} & & \downarrow{\scriptstyle (u_1,\dots,u_\ell)^*} \\ End(X)(v; v_1,\dots,v_q) & \xrightarrow[()_{\bar f}]{} & End(X)(v_{\bar f}; (v_1)_{\bar f},\dots,(v_q)_{\bar f}) \end{array}$$

*commutes, where* $v = \gamma^{\times k}(w; u_{f1},\dots,u_{fm})$ *and* $v_{\bar f} = \gamma^{\times k}(w_f; u_1,\dots,u_\ell)$ *etc.*

(14) *End(I)-substitution and End(I)-composition commute.*
*Let* $w, w_i, w_{ij} \in Mor_{End(I)^k}(m,1)$ *and* $u_i \in End(I)(k_i)$ *for* $i = 1,\dots,m$. *Let* $\alpha \in End(X)(w; w_1,\dots,w_q)$, $\alpha_\ell \in End(X)(w_\ell; w_{\ell 1},\dots,w_{\ell p_\ell})$ *for* $\ell = 1,\dots,q$ *and* $\beta := (u_1,\dots,u_m)^*\alpha$ *etc. Then*

$$(u_1,\dots,u_m)^*\gamma(\alpha;\alpha_1,\dots,\alpha_q) = \gamma(\beta;\beta_1,\dots,\beta_q).$$

This concludes our motivational discussion of the 2-theory $End(X)$ fibered over the theory $End(I)$ for a 2-functor $X : I^2 \to Cat$. Next we turn to the general discussion.

## 13.2. 2-Theories and Algebras over 2-Theories

A general 2-theory has all of the properties described in the example above. P. Hu and I. Kriz introduce the notion of a 2-theory in [**25**] as follows.

DEFINITION 13.3. A *2-theory* $\Theta$ *fibered over the theory* $T$, written $(\Theta, T)$ for short, is a natural number $k$, a theory $T$, and a contravariant functor $\Theta : T \to Cat$ from the category $T$ to the 2-category $Cat$ of small categories such that

- $Obj\ \Theta(m) = \coprod_{n\geq 0} Mor_{T^k}(m,n)$ for all $m \in \mathbb{N}$, where $T^k$ is the theory with the same objects as $T$, but with $Mor_{T^k}(m,n) = Mor_T(m,n)^k$
- If $w_1,\dots,w_n \in Mor_{T^k}(m,1)$, then the word in $Mor_{T^k}(m,n)$ with which the $n$-tuple $w_1,\dots,w_n$ is identified is the product in $\Theta(m)$ of $w_1,\dots,w_n$

- For $w \in Mor_T(m, n)$ the functor $\Theta(w) : \Theta(n) \to \Theta(m)$ is $\Theta(w)(v) = v \circ w^{\times k}$ on objects $v \in Mor_{T^k}(n, j)$.

For objects $w_1, \ldots, w_n, w \in Mor_{T^k}(m, 1) \subseteq Obj\, \Theta(m)$ we set

$$\Theta(w; w_1, \ldots, w_n) := Mor_{\Theta(m)}(\prod_{i=1}^{n} w_i, w).$$

The second condition explains the choice of notation $\prod_{i=1}^{n} w_i$. Given a 2-theory such as this, it has operations and relations as in Theorem 13.1. Vice-a-versa, given sets $\Theta(w; w_1, \ldots, w_n) := Mor_{\Theta(m)}(\prod_{i=1}^{n} w_i, w)$ with operations and relations as in Theorems 13.1 and 13.2 we get a 2-theory. We refer to these operations and relations as the *operations and relations of 2-theories*. Recall that a pseudo algebra $I$ over a theory $T$ is a category such that for every word $w \in T(n)$ we have a functor $\Phi_n(w) : I^n \to I$. Moreover, for every operation of theories (composition, substitution, and identity) we have a coherence iso and for every relation of theories we have a coherence diagram. A pseudo $(\Theta, T)$-algebra can be defined analogously.

DEFINITION 13.4. Let $(\Theta, T)$ be a 2-theory. A *pseudo $(\Theta, T)$-algebra over $I^k$* consists of the following data:
- a small pseudo $T$-algebra $I$ with structure maps $\Phi : T(n) \to Functors(I^n, I)$
- a strict 2-functor $X : I^k \to Cat$
- set maps $\phi : \Theta(w; w_1, \ldots, w_n) \to End(X)(\Phi(w); \Phi(w_1), \ldots, \Phi(w_n))$, where $\Phi(w)$ means to apply $\Phi$ to each component of $w$ to make $I^k$ into the product pseudo $T$-algebra of $k$ copies of $I$
- a coherence iso modification for each operation of 2-theories and these coherence iso modifications satisfy coherence diagrams indexed by the relations of 2-theories.

A morphism of pseudo $(\Theta, T)$-algebras over $I^k$ is similar to a morphism of pseudo $T$-algebras.

DEFINITION 13.5. Let $X, Y : I^k \to Cat$ be pseudo $(\Theta, T)$-algebras over $I^k$. Then a *morphism $H : X \to Y$ of pseudo $(\Theta, T)$-algebras over $I^k$* is a strict 2-natural transformation $H : X \Rightarrow Y$ with coherence iso modifications $\rho_\alpha$ indexed by elements $\alpha \in \Theta(w; w_1, \ldots, w_n)$, where $w, w_1, \ldots, w_n \in Obj\, \Theta(m)$.

$$\begin{array}{ccc}
X \circ \Phi(w_1) \circ d^m \times \cdots \times X \circ \Phi(w_n) \circ d^m & \xrightarrow{\phi_X(\alpha)} & X \circ \Phi(w) \circ d^m \\
{\scriptstyle H * i_{\Phi(w_1)} * i_{d^m}} \Big\Downarrow \quad \cdots \quad \Big\Downarrow {\scriptstyle H * i_{\Phi(w_n)} * i_{d^m}} & \overset{\rho_\alpha}{\rightsquigarrow} & \Big\Downarrow {\scriptstyle H * i_{\Phi(w)} * i_{d^m}} \\
Y \circ \Phi(w_1) \circ d^m \times \cdots \times Y \circ \Phi(w_n) \circ d^m & \xrightarrow[\phi_Y(\alpha)]{} & Y \circ \Phi(w) \circ d^m
\end{array}$$

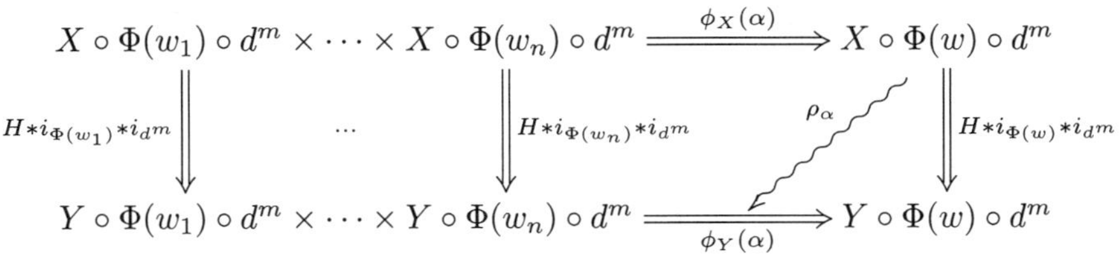

The coherence iso modification $\rho_\alpha$ is required to commute with all coherence iso modifications of the pseudo algebra structure.

The 2-cells of pseudo $(\Theta, T)$-algebras over $I^k$ are also similar to the 2-cells of pseudo $T$-algebras.

DEFINITION 13.6. Let $G, H : X \to Y$ be morphisms of pseudo $(\Theta, T)$-algebras over $I^k$. Then a *2-cell $\sigma : G \Rightarrow H$* is a modification which commutes with the coherence iso modifications $\rho^G$ and $\rho^H$ appropriately.

THEOREM 13.7. *The pseudo $(\Theta, T)$-algebras over $I^k$ form a 2-category.*

*Proof:* Routine. □

## 13.3. The Algebraic Structure of Rigged Surfaces

The purpose of this section is to introduce the category of rigged surfaces as an example of a pseudo algebra over a 2-theory fibered over a theory and to describe its stack structure. This approach was introduced in [**25**] by P. Hu and I. Kriz. In their terminology, a smooth, compact, not necessarily connected, 2-dimensional manifold $x$ with a complex structure is called a *rigged surface* if each boundary component $k$ comes equipped with a parametrization diffeomorphism $f_k : S^1 \to k$ which is analytic with respect to the complex structure on $x$, *i.e.* the diffeomorphism $f_k$ extends to a holomorphic map when we go into local coordinates. A boundary component $k$ is called *inbound* or *outbound* depending on the orientation of its parametrization $f_k$ with respect to the orientation on $k$ induced by the complex structure. The convention is to call the identity parametrization of the boundary of the unit disk *inbound.* A *morphism* of rigged surfaces is a holomorphic diffeomorphism which preserves the boundary parametrizations.

The structure of the category of rigged surfaces has the following features, which were studied in [**25**]. For finite sets $a$ and $b$, let $Obj\ X_{a,b}$ denote the set of rigged surfaces $x$ equipped with a bijection between the inbound boundary components of $x$ and $a$ as well as a bijection between the outbound boundary components of $x$ and $b$. For $x, y \in Obj\ X_{a,b}$, let $Mor_{X_{a,b}}(x,y)$ be the morphisms of rigged surfaces which preserve the bijections with $a$ and $b$. For finite sets $a, b, c,$ and $d$ we can take the disjoint union of any two rigged surfaces $x \in Obj\ X_{a,b}$ and $y \in Obj\ X_{c,d}$ and the result is an element of $Obj\ X_{a \coprod c, b \coprod d}$. We can apply this process to morphisms as well, and we get a functor $\coprod : X_{a,b} \times X_{c,d} \to X_{a \coprod c, b \coprod d}$ called *disjoint union.* Note that this functor is indexed by the finite sets $a, b, c,$ and $d$. For finite sets $a, b,$ and $c$ we also have a *gluing functor* $\check{?} : X_{a \coprod c, b \coprod c} \to X_{a,b}$ which identifies an inbound boundary component $k$ with an outbound boundary component $k'$ according to $f_k(z) \sim f_{k'}(z)$ for all $z \in S^1$ whenever $k$ and $k'$ are labelled by the same element of $c$. This gluing functor is also indexed by the finite sets $a, b,$ and $c$. There is also a *unit* 0 in $X_{0,0}$ given by the empty set. These disjoint union functors, gluing functors, and unit along with their coherence isos and coherence diagrams give the category of rigged surfaces the structure of a *pseudo algebra over the 2-theory of commutative monoids with cancellation.* More precisely, if $I$ denotes the category of finite sets and bijections, then the assignment $(a, b) \mapsto X_{a,b}$ defines a strict 2-functor $X : I^2 \to Cat$ which is a pseudo algebra over the 2-theory which we now describe.

We define the 2-theory $(\Theta, T)$ of *commutative monoids with cancellation* as follows. Let $T$ be the theory of commutative monoids and let $+ : 2 \to 1$ and $0 : 0 \to 1$ be the usual words in the theory of commutative monoids. Let $k = 2$. The 2-theory $\Theta$ is generated by three words: *addition* $+$, *cancellation* $\check{?}$, and *unit* 0. These are described in terms of a general algebra $X : I^2 \to Sets$ over $(\Theta, T)$ as follows. Note that $+$ and 0 have two meanings.

$$+ : X_{a,b} \times X_{c,d} \to X_{a+c,b+d}$$

$$\check{?} : X_{a+c,b+c} \to X_{a,b}$$

$$0 \in X_{0,0}$$

These generating words must satisfy the following axioms.

(1) The word + is *commutative*.

$$\begin{CD}
X_{a,b} \times X_{c,d} @>+>> X_{a+c,b+d} \\
@VVV @| \\
X_{c,d} \times X_{a,b} @>>+> X_{c+a,d+b}
\end{CD}$$

(2) The word + is *associative*.

$$\begin{CD}
(X_{a,b} \times X_{c,d}) \times X_{e,f} @>+\times 1_{X_{e,f}}>> X_{a+c,b+d} \times X_{e,f} \\
@VVV @VV+V \\
X_{a,b} \times (X_{c,d} \times X_{e,f}) @. X_{(a+c)+e,(b+d)+f} \\
@V1_{X_{a,b}}\times +VV @| \\
X_{a,b} \times X_{c+e,d+f} @>>+> X_{a+(c+e),b+(d+f)}
\end{CD}$$

(3) The word + has *unit* $0 \in X_{0,0}$.

$$\begin{array}{ccc}
X_{a,b} \times \{0\} & \xrightarrow{\;+\;} & X_{a+0,b+0} \\
& \searrow^{pr_1} & \| \\
& & X_{a,b}
\end{array}$$

(4) The word $\check{?}$ is *transitive*.

$$\begin{CD}
X_{(a+c)+d,(b+c)+d} @>\check{?}>> X_{a+c,b+c} \\
@| @VV\check{?}V \\
X_{a+(c+d),b+(c+d)} @>>\check{?}> X_{a,b}
\end{CD}$$

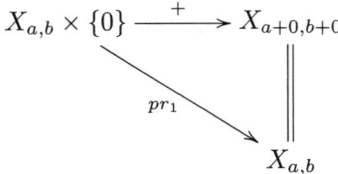

(5) The word $\check{?}$ *distributes* over the word $+$.

$$\begin{array}{ccc}
X_{a+c,b+c} \times X_{e,f} & \xrightarrow{+} & X_{(a+c)+e,(b+c)+f} \\
 & & \| \\
\downarrow{\scriptstyle \check{?} \times 1_{X_{e,f}}} & & X_{(a+e)+c,(b+f)+c} \\
 & & \downarrow{\scriptstyle \check{?}} \\
X_{a,b} \times X_{e,f} & \xrightarrow[+]{} & X_{a+e,b+f}
\end{array}$$

(6) Trivial cancellation is trivial.

$$\begin{array}{ccc}
X_{a+0,b+0} & \xrightarrow{\check{?}} & X_{a,b} \\
 & \diagdown\!\!\diagdown & \downarrow{\scriptstyle 1_{X_{a,b}}} \\
 & & X_{a,b}
\end{array}$$

The category of rigged surfaces forms a pseudo algebra over this 2-theory of commutative monoids with cancellation. The category $I$ of finite sets and bijections equipped with the operation $\coprod$ is a pseudo algebra over the theory $T$ of commutative monoids. The pseudo algebra structure on $X : I^2 \to Cat$ is given by assigning a fixed choice of $\coprod$ to $+$, gluing of manifolds to $\check{?}$, and the empty set to 0. This defines the structure maps $\Theta(w; w_1, \ldots, w_n) \to End(X)(w; w_1, \ldots, w_n)$.

In [**25**] and [**26**] the algebraic structure of holomorphic families of rigged surfaces is captured by a stack of pseudo algebras over the 2-theory of commutative monoids with cancellation, which is also called a stack of lax commutative monoids with cancellation (SLCMC). We describe this stack now. Let $\mathcal{B}$ be the category of finite dimensional complex manifolds with morphisms holomorphic maps. A collection $\{B_i \to B\}_i$ of (open) holomorphic embeddings are a cover if their combined image covers $B$. This makes $\mathcal{B}$ into a Grothendieck topology. For any finite dimensional complex manifold $B$ let $I^B$ denote the category of covering spaces of $B$ with finite fibers and morphisms given by isomorphisms of covering spaces. The category $I^B$ is a pseudo commutative monoid under $\coprod$. Let $s$ and $t$ be objects of $\mathcal{B}$. Define $X^B_{s,t}$ as the category of holomorphic families of rigged surfaces over $B$ with inbound boundary components labelled by the covering space $s$ of $B$ and outbound boundary components labelled by the covering space $t$ of $B$. Such a holomorphic family $x$ is by definition a complex manifold $x$ with analytic boundary and a transverse holomorphic map $p : x \to B$ such that $x_b = p^{-1}(b)$ is a rigged surface for all $b \in B$. Moreover, the boundary parametrizations of $p^{-1}(b)$ vary holomorphically with $b$ in the precise sense on page 330 of [**25**]. To say that the inbound boundary components of $x$ are labelled by the covering space $s$ means that for each $b \in B$ the rigged surface $x_b$ is equipped with a bijection between its inbound boundary components and the fiber of $s$ over $b$. The explanation for the covering space $t$ labelling the outbound boundary components is similar. With these definitions as well as disjoint union, gluing, and empty set, the functor $X^B : (I^B)^2 \to Cat$ is a pseudo algebra over the 2-theory of commutative monoids with cancellation.

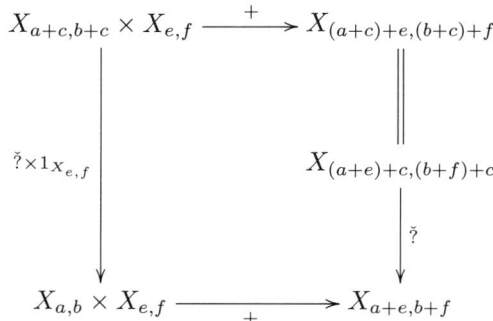

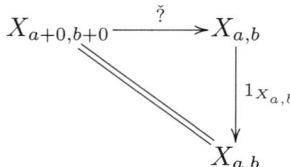

Let $\mathcal{C}$ denote the 2-category of pseudo algebras over the 2-theory of commutative monoids with cancellation. This 2-category admits bilimits, which we prove in a special case in the next section. Define a contravariant pseudo functor $G : \mathcal{B} \longrightarrow \mathcal{C}$ by taking a finite dimensional complex manifold $B$ to the pseudo algebra $X^B$ over the 2-theory of commutative monoids with cancellation with underlying pseudo commutative monoid $I^B$. Then $G$ takes Grothendieck covers to bilimits because it does so on the underlying categories comprising the pseudo algebras. Hence $G$ is a stack. It is in this sense that the category of rigged surfaces forms a stack.

## 13.4. Weighted Pseudo Limits of Pseudo $(\Theta, T)$-Algebras

The 2-category of pseudo $(\Theta, T)$-algebras admits weighted pseudo limits, just like the 2-category of pseudo $T$-algebras. In the following theorem we prove this for pseudo $(\Theta, T)$-algebras with fixed underlying pseudo $T$-algebra $I^k$. The proof can be modified to the general case of pseudo $(\Theta, T)$-algebras with different underlying pseudo $T$-algebras by taking the pseudo limit of the underlying pseudo $T$-algebras as well.

THEOREM 13.8. *Let $\mathcal{J}$ be a 1-category and $\mathcal{C}$ the 2-category of pseudo $(\Theta, T)$-algebras over $I^k$. Let $F : \mathcal{J} \longrightarrow \mathcal{C}$ be a pseudo functor. Then $F$ admits a pseudo limit $(X, \pi)$ in $\mathcal{C}$, where $\pi : \Delta_X \Rightarrow F$ is a universal pseudo cone.*

*Proof:* Let $\gamma$ and $\delta$ be the 2-cells in $\mathcal{C}$ which make $F$ into a pseudo functor. For each $j \in Obj\ \mathcal{J}$, let $X^j : I^k \longrightarrow Cat$ be the strict 2-functor belonging to the pseudo $(\Theta, T)$-algebra $Fj$. Then for each fixed object $i \in Obj\ I^k$ and each object $j \in \mathcal{J}$ we have a category $X_i^j$. For each morphism $f : j \longrightarrow m$ in $\mathcal{J}$, the map $Ff : X^j \Rightarrow X^m$ is a strict 2-natural transformation which gives us a functor $(Ff)_i : X_i^j \longrightarrow X_i^m$ for each $i \in Obj\ I^k$. Thus for fixed $i$ we have a pseudo functor $F_i : \mathcal{J} \longrightarrow Cat$ defined by $j \mapsto X_i^j$ and $f \mapsto (Ff)_i$. The coherence isos of $F_i$ are the coherence iso modifications of $F$ evaluated at $i$.

Let $X_i := PseudoCone(\mathbf{1}, F_i)$, where $\mathbf{1}$ is the terminal object in the category of small categories. Then it is known from Chapter 5 that $X_i$ is the pseudo limit of $F_i$ in $Cat$. Proceeding analogously on morphisms of $I^k$, we obtain a strict 2-functor $X : I^k \longrightarrow Cat$ defined by $i \mapsto X_i$. More precisely, if $h : i_1 \longrightarrow i_2$ is a morphism in $I^k$ and $\eta \in Obj\ X_{i_1}$, then $X_h(\eta)(j) := X_h^j(\eta(j))$ for $j \in Obj\ \mathcal{J}$.

A more conceptual way to view the construction of the strict 2-functor $X : I^k \longrightarrow Cat$ is the following. For $i \in I^k$, let $F_i : \mathcal{J} \longrightarrow Cat$ be the pseudo functor from above. For a morphism $h : i_1 \longrightarrow i_2$ in $I^k$, let $F_h : F_{i_1} \Rightarrow F_{i_2}$ be the pseudo natural transformation given by $F_h(j) := X_h^j$. The pseudo natural transformation $F_h$ is actually strictly 2-natural because $Ff : X^j \Rightarrow X^m$ is a strict 2-natural transformation for each $f : j \longrightarrow m$ in $\mathcal{J}$. Thus $i \mapsto F_i$ and $h \mapsto F_h$ define a strict functor $I^k \longrightarrow Functors(\mathcal{J}, Cat)$. Now recall that $PseudoCone(\mathbf{1}, -)$ is a covariant functor from $Functors(\mathcal{J}, Cat)$ to $Cat$. The composition

$$I^k \longrightarrow Functors(\mathcal{J}, Cat) \xrightarrow{PseudoCone(\mathbf{1},-)} Cat$$

is $X : I^k \longrightarrow Cat$.

We claim that this 2-functor $X : I^k \longrightarrow Cat$ has the structure of a pseudo $(\Theta, T)$-algebra. The argument is like Lemma 8.2, although the coherences need some care. First we define maps $\phi : \Theta(w; w_1, \ldots, w_n) \longrightarrow End(X)(\Phi(w); \Phi(w_1), \ldots, \Phi(w_n))$, where $w_1, \ldots, w_n, w \in Mor_{T^k}(m,1)$. Let $\alpha \in \Theta(w; w_1, \ldots, w_n)$. We need to define a natural transformation

$$\phi(\alpha) : X \circ \Phi(w_1) \circ d^m \times \cdots \times X \circ \Phi(w_n) \circ d^m \Rightarrow X \circ \Phi(w) \circ d^m$$

"componentwise," where $d^m : I^m \to (I^m)^k$ is the diagonal functor. Let

$$\phi_j : \Theta(w; w_1, \ldots, w_n) \longrightarrow End(X^j)(\Phi(w); \Phi(w_1), \ldots, \Phi(w_n))$$

be the maps that make $X^j : I^k \to Cat$ into a pseudo $(\Theta, T)$-algebra for each $j \in Obj\ \mathcal{J}$. Let $i \in I^m$. We define a functor

$$(\phi(\alpha))_i : X_{\Phi(w_1)\circ d^m(i)} \times \cdots \times X_{\Phi(w_n)\circ d^m(i)} \longrightarrow X_{\Phi(w)\circ d^m(i)}$$

and show that $i \mapsto (\phi(\alpha))_i$ is natural. Recall that objects of

$$X_{\Phi(w_\ell)\circ d^m(i)} = PseudoCone(\mathbf{1}, F_{\Phi(w_\ell)\circ d^m(i)})$$

can be identified with a subset of

$$\{(a_j)_j \times (\varepsilon_f)_f \in \prod_{j \in Obj\ \mathcal{J}} Obj\ X^j_{\Phi(w_\ell)\circ d^m(i)} \times \prod_{f \in Mor\ \mathcal{J}} Mor\ X^{Tf}_{\Phi(w_\ell)\circ d^m(i)} |$$

$$\varepsilon_f : (Ff)_{\Phi(w_\ell)\circ d^m(i)}(a_{Sf}) \to a_{Tf} \text{ is iso for all } f \in Mor\ \mathcal{J}\}$$

by Remark 5.4. A similar statement holds for morphisms according to 5.5. Let $\eta^\ell = (a_j^\ell)_j \times (\varepsilon_f^\ell)_f \in Obj\ X_{\Phi(w_\ell)\circ d^m(i)}$ and $(\xi_j^\ell)_j \in Mor\ X_{\Phi(w_\ell)\circ d^m(i)}$ for $1 \leq \ell \leq n$. Define

$$a_j := (\phi_j(\alpha))_i(a_j^1, \ldots, a_j^n)$$

and

$$\varepsilon_f := (\phi_{Tf}(\alpha))_i(\varepsilon_f^1, \ldots, \varepsilon_f^n) \circ (\rho_\alpha^{Ff})_i(a_{Sf}^1, \ldots, a_{Sf}^n).$$

Note that

$$(\rho_\alpha^{Ff})_i(a_{Sf}^1, \ldots, a_{Sf}^n) : (Ff)_{\Phi(w_\ell)\circ d^m(i)}(\phi_{Sf}(\alpha))_i(a_{Sf}^1, \ldots, a_{Sf}^n) \longrightarrow$$
$$(\phi_{Tf}(\alpha))_i((Ff)_{\Phi(w_\ell)\circ d^m(i)}(a_{Sf}^1), \ldots, (Ff)_{\Phi(w_\ell)\circ d^m(i)}(a_{Sf}^n))$$

and the composition in the definition of $\varepsilon_f$ makes sense. Also define

$$\xi_j := (\phi_j(\alpha))_i(\xi_j^1, \ldots, \xi_j^n).$$

Then $\phi(\alpha)$ is defined "componentwise" by

$$(\phi(\alpha))_i(\eta^1, \ldots, \eta^n) := (a_j)_j \times (\varepsilon_f)$$

and

$$(\phi(\alpha))_i((\xi_j^1)_j, \ldots, (\xi_j^n)_j) := (\xi_j)_j.$$

By an argument similar to the proof of Lemma 8.2, these images are actually in $X_{\Phi(w)\circ d^m(i)}$. Next note that $i \mapsto (\phi(\alpha))_i$ is natural because $i \mapsto (\phi_j(\alpha))_i$ is natural for all $j \in Obj\ \mathcal{J}$, *i.e.* $i \mapsto (\phi(\alpha))_i$ is natural in each "coordinate" and is therefore natural. Hence we have constructed set maps $\phi : \Theta(w; w_1, \ldots, w_n) \to End(X)(\Phi(w); \Phi(w_1), \ldots, \Phi(w_n))$.

We define the coherence iso modifications for $\phi$ to be those modifications which have the coherence iso modifications for $\phi_j$ in the $j$-th coordinate. For example, we define the identity modification $I_w : 1_{\Phi(w)} \rightsquigarrow \phi(1_w)$ by

$$I_w((a_j)_j \times (\varepsilon_f)_f) := (I_w^j(a_j))_j$$

for $i \in I^m$ and $(a_j)_j \times (\varepsilon_f)_f \in X_{\Phi(w)\circ d^m(i)}$. The arrow $I_w((a_j)_j \times (\varepsilon_f)_f)$ is an arrow in the category $X_{\Phi(w)\circ d^m(i)}$ by an argument like the proof of Lemma 8.2.

Similarly, we can show that these assignments are modifications and that the coherence diagrams are satisfied because everything is done componentwise. Hence $X : I^k \to Cat$ has the structure of a pseudo $(\Theta, T)$-algebra.

Next we need a universal pseudo cone $\pi : \Delta_X \Rightarrow F$, where $\Delta_X : \mathcal{J} \to \mathcal{C}$ is the constant functor which takes everything to $X$. Define a natural transformation $\pi_j : X \Rightarrow X^j$ by letting $\pi_j(i) : X_i \Rightarrow X_i^j$ be the projection. The natural transformation $\pi_j$ commutes with the $(\Theta, T)$ structure maps, and so $\pi_j$ is a morphism of pseudo $(\Theta, T)$-algebras by taking the coherence iso modifications to be trivial. The assignment $j \mapsto \pi_j$ is pseudo natural with coherence 2-cell $\tau_{j,m}(f) : Ff \circ \pi_j \Rightarrow \pi_m$ for each $f : j \to m$ in $\mathcal{J}$ as in the 1-theory case. A similar argument to the 1-theory case shows that $\tau_{j,m}(f)$ is a 2-cell in $\mathcal{C}$. Hence, we have a pseudo natural transformation $\pi : \Delta_X \Rightarrow F$. We can prove the universality of $\pi$ by applying the argument in the lemmas leading up to Theorem 8.9 to $X_i \to X_i^j$ for each fixed $i \in Obj\ I^k$ and then passing to functors $I^k \to Cat$. We must of course take the coherence isos into consideration.

We conclude that $(X, \pi)$ is a pseudo limit of the pseudo functor $F : \mathcal{J} \to \mathcal{C}$. □

THEOREM 13.9. *The 2-category of pseudo $(\Theta, T)$-algebras over $I^k$ admits pseudo limits.*

*Proof:* This follows immediately from the previous theorem. □

LEMMA 13.10. *The 2-category $\mathcal{C}$ of pseudo $(\Theta, T)$-algebras admits cotensor products.*

*Proof:* Let $J \in Obj\ Cat$ and let $F : I^k \to Cat$ be a pseudo $(\Theta, T)$-algebra. Define a strict 2-functor $P : I^k \to Cat$ by $P_i := (F_i)^J$, which is the 1-category of 1-functors $J \to F_i$. We claim that $P$ has the structure of a pseudo $(\Theta, T)$-algebra. This structure is obtained by doing the operations pointwise. Let $\phi : \Theta(w; w_1, \ldots, w_n) \to End(F)(\Phi(w); \Phi(w_1), \ldots, \Phi(w_n))$ denote the maps which make $F$ into a pseudo $(\Theta, T)$-algebra. Then define

$$\phi^P : \Theta(w; w_1, \ldots, w_n) \to End(P)(\Phi(w); \Phi(w_1), \ldots, \Phi(w_n))$$
$$\phi^P(\alpha)_i(\eta^1, \ldots, \eta^n)(j) := \phi(\alpha)_i(\eta^1(j), \ldots, \eta^n(j))$$

for functors $\eta^\ell : J \to X_{\Phi(w_\ell) \circ d^m(i)}$ with $1 \leq \ell \leq n$. Coherence isos can also be defined in this manner. Then the coherence diagrams commute because they commute pointwise. Hence $P$ is a pseudo $(\Theta, T)$-algebra.

A proof similar to the proof of Lemma 8.11 shows that $P$ is the cotensor product of $J$ and $F$. We must apply the argument for $F$ in Lemma 8.11 to each $F_i$ for $i \in Obj\ I^k$. □

THEOREM 13.11. *The 2-category $\mathcal{C}$ of pseudo $(\Theta, T)$-algebras admits weighted pseudo limits.*

*Proof:* By Theorem 13.9 it admits pseudo limits, and hence it admits pseudo equalizers. The 2-category $\mathcal{C}$ obviously admits products. By Lemma 13.10 it admits cotensor products. Hence by Theorem 3.22 it admits weighted pseudo limits. □

THEOREM 13.12. *The 2-category $\mathcal{C}$ of pseudo $(\Theta, T)$-algebras admits weighted bilimits.*

*Proof:* The 2-category $\mathcal{C}$ admits weighted pseudo limits, so it also admits weighted bilimits. □

# Bibliography


[1] J. Adámek, F. W. Lawvere, and J. Rosický, *How algebraic is algebra?*, Theory Appl. Categ. **8** (2001), 253–283 (electronic). MR1825435 (2002b:18005)

[2] ———, *Continuous categories revisited*, Theory Appl. Categ. **11** (2003), No. 11, 252–282 (electronic). MR1988399 (2004e:18002)

[3] ———, *On the duality between varieties and algebraic theories*, Algebra Universalis **49** (2003), no. 1, 35–49. MR1978611 (2004b:18010)

[4] John C. Baez and James Dolan, *Higher-dimensional algebra and topological quantum field theory*, J. Math. Phys. **36** (1995), no. 11, 6073–6105. MR1355899 (97f:18003)

[5] ———, *Categorification*, Higher category theory (Evanston, IL, 1997), Contemp. Math., vol. 230, Amer. Math. Soc., Providence, RI, 1998, pp. 1–36. MR1664990 (99k:18016)

[6] Jean Bénabou, *Introduction to bicategories*, Reports of the Midwest Category Seminar, Springer, Berlin, 1967, pp. 1–77. MR0220789 (36 #3841)

[7] ———, *Structures algébriques dans les catégories*, Cahiers Topologie Géom. Différentielle **10** (1968), 1–126. MR0244335 (39 #5650)

[8] G. J. Bird, G. M. Kelly, A. J. Power, and R. H. Street, *Flexible limits for 2-categories*, J. Pure Appl. Algebra **61** (1989), no. 1, 1–27. MR1023741 (91a:18009)

[9] R. Blackwell, G. M. Kelly, and A. J. Power, *Two-dimensional monad theory*, J. Pure Appl. Algebra **59** (1989), no. 1, 1–41. MR1007911 (91a:18010)

[10] Francis Borceux, *Handbook of categorical algebra. 1*, Encyclopedia of Mathematics and its Applications, vol. 50, Cambridge University Press, Cambridge, 1994, Basic category theory. MR1291599 (96g:18001a)

[11] Francis Borceux and G. M. Kelly, *A notion of limit for enriched categories*, Bull. Austral. Math. Soc. **12** (1975), 49–72. MR0369477 (51 #5710)

[12] Richard E. Borcherds, *Monstrous moonshine and monstrous Lie superalgebras*, Invent. Math. **109** (1992), no. 2, 405–444. MR1172696 (94f:11030)

[13] Lawrence Breen, *On the classification of 2-gerbes and 2-stacks*, Astérisque (1994), no. 225, 160. MR1301844 (95m:18006)

[14] Aurelio Carboni, Scott Johnson, Ross Street, and Dominic Verity, *Modulated bicategories*, J. Pure Appl. Algebra **94** (1994), no. 3, 229–282. MR1285544 (96f:18008)

[15] Gerald Dunn, *Lax operad actions and coherence for monoidal $N$-categories, $A_\infty$ rings and modules*, Theory Appl. Categ. **3** (1997), No. 4, 50–84 (electronic). MR1432192 (97i:18010)

[16] Charles Ehresmann, *Catégories structurées*, Ann. Sci. École Norm. Sup. (3) **80** (1963), 349–426. MR0197529 (33 #5694)

[17] Barbara Fantechi, *Stacks for everybody*, European Congress of Mathematics, Vol. I (Barcelona, 2000), Progr. Math., vol. 201, Birkhäuser, Basel, 2001, pp. 349–359. MR1905329 (2003h:14003)

[18] Jean Giraud, *Cohomologie non abélienne*, Springer-Verlag, Berlin, 1971, Die Grundlehren der mathematischen Wissenschaften, Band 179. MR0344253 (49 #8992)

[19] John W. Gray, *Formal category theory: adjointness for 2-categories*, Springer-Verlag, Berlin, 1974, Lecture Notes in Mathematics, Vol. 391. MR0371990 (51 #8207)

[20] ———, *Quasi-Kan extensions for 2-categories*, Bull. Amer. Math. Soc. **80** (1974), 142–147. MR0340369 (49 #5124)

[21] ———, *Closed categories, lax limits and homotopy limits*, J. Pure Appl. Algebra **19** (1980), 127–158. MR593251 (82f:18007a)

[22] ———, *The existence and construction of lax limits*, Cahiers Topologie Géom. Différentielle **21** (1980), no. 3, 277–304. MR591387 (82f:18007b)

[23] ———, *The representation of limits, lax limits and homotopy limits as sections*, Mathematical applications of category theory (Denver, Col., 1983), Contemp. Math., vol. 30, Amer. Math. Soc., Providence, RI, 1984, pp. 63–83. MR749769 (85k:18011)

[24] Claudio Hermida, *From coherent structures to universal properties*, J. Pure Appl. Algebra **165** (2001), no. 1, 7–61. MR1860877 (2002g:18008)

[25] P. Hu and I. Kriz, *Conformal field theory and elliptic cohomology*, Adv. Math. **189** (2004), no. 2, 325–412, `http://www.math.lsa.umich.edu/~ikriz/`. MR2101224

[26] ———, *Closed and open conformal field theories and their anomalies*, Comm. Math. Phys. **254** (2005), no. 1, 221–253, `http://www.math.lsa.umich.edu/~ikriz/`. MR2116744

[27] P. Hu, I. Kriz, and A. Voronov, *On Kontsevich's Hochschild cohomology conjecture*, to appear, `http://www.math.lsa.umich.edu/~ikriz/`.

[28] G. M. Kelly, *Basic concepts of enriched category theory*, London Mathematical Society Lecture Note Series, vol. 64, Cambridge University Press, Cambridge, 1982. MR651714 (84e:18001)

[29] ———, *Elementary observations on* 2*-categorical limits*, Bull. Austral. Math. Soc. **39** (1989), no. 2, 301–317. MR998024 (90f:18004)

[30] G. M. Kelly and A. J. Power, *Adjunctions whose counits are coequalizers, and presentations of finitary enriched monads*, J. Pure Appl. Algebra **89** (1993), no. 1-2, 163–179. MR1239558 (94k:18008)

[31] G. M. Kelly and Ross Street, *Review of the elements of* 2*-categories*, Category Seminar (Proc. Sem., Sydney, 1972/1973), Springer, Berlin, 1974, pp. 75–103. Lecture Notes in Math., Vol. 420. MR0357542 (50 #10010)

[32] Stephen Lack, *On the monadicity of finitary monads*, J. Pure Appl. Algebra **140** (1999), no. 1, 65–73. MR1700570 (2000g:18004)

[33] ———, *Codescent objects and coherence*, J. Pure Appl. Algebra **175** (2002), no. 1-3, 223–241, Special volume celebrating the 70th birthday of Professor Max Kelly. MR1935980 (2003k:18008)

[34] F. William Lawvere, *Functorial semantics of algebraic theories*, Proc. Nat. Acad. Sci. U.S.A. **50** (1963), 869–872. MR0158921 (28 #2143)

[35] ———, *Some algebraic problems in the context of functorial semantics of algebraic theories*, Reports of the Midwest Category Seminar. II, Springer, Berlin, 1968, pp. 41–61. MR0231882 (38 #210)

[36] ———, *Ordinal sums and equational doctrines*, Sem. on Triples and Categorical Homology Theory (ETH, Zürich, 1966/67), Springer, Berlin, 1969, pp. 141–155. MR0240158 (39 #1512)

[37] Saunders Mac Lane, *Natural associativity and commutativity*, Rice Univ. Studies **49** (1963), no. 4, 28–46. MR0170925 (30 #1160)

[38] ———, *Coherence theorems and conformal field theory*, Category theory 1991 (Montreal, PQ, 1991), CMS Conf. Proc., vol. 13, Amer. Math. Soc., Providence, RI, 1992, pp. 321–328. MR1192155 (94d:18010)

[39] ———, *Categories for the working mathematician*, second ed., Graduate Texts in Mathematics, vol. 5, Springer-Verlag, New York, 1998. MR1712872 (2001j:18001)

[40] Saunders Mac Lane and Ieke Moerdijk, *Sheaves in geometry and logic*, Universitext, Springer-Verlag, New York, 1994, A first introduction to topos theory, Corrected reprint of the 1992 edition. MR1300636 (96c:03119)

[41] John L. MacDonald and Arthur Stone, *Soft adjunction between 2-categories*, J. Pure Appl. Algebra **60** (1989), no. 2, 155–203. MR1020715 (90i:18004)

[42] Ieke Moerdijk, *Introduction to the language of stacks and gerbes*, `http://arxiv.org/abs/math.AT/0212266`.

[43] Joseph Polchinski, *String theory. Vol. I*, Cambridge Monographs on Mathematical Physics, Cambridge University Press, Cambridge, 1998, An introduction to the bosonic string. MR1648555 (99h:81183)

[44] A. J. Power, *Enriched Lawvere theories*, Theory Appl. Categ. **6** (1999), 83–93 (electronic), The Lambek Festschrift. MR1732465 (2000j:18002)

[45] Graeme Segal, *The definition of conformal field theory*, Topology, geometry and quantum field theory, London Math. Soc. Lecture Note Ser., vol. 308, Cambridge Univ. Press, Cambridge, 2004, pp. 421–577. MR2079383

[46] Harold Simmons, *The glueing construction and lax limits*, Math. Structures Comput. Sci. **4** (1994), no. 4, 393–431. MR1322182 (96d:18002)

[47] Ross Street, *The formal theory of monads*, J. Pure Appl. Algebra **2** (1972), no. 2, 149–168. MR0299653 (45 #8701)

[48] ———, *Two constructions on lax functors*, Cahiers Topologie Géom. Différentielle **13** (1972), 217–264. MR0347936 (50 #436)

[49] ———, *Limits indexed by category-valued 2-functors*, J. Pure Appl. Algebra **8** (1976), no. 2, 149–181. MR0401868 (53 #5695)

[50] ———, *Fibrations in bicategories*, Cahiers Topologie Géom. Différentielle **21** (1980), no. 2, 111–160. MR574662 (81f:18028)

[51] ———, *Correction to: "Fibrations in bicategories"*, Cahiers Topologie Géom. Différentielle Catég. **28** (1987), no. 1, 53–56. MR903151 (88i:18004)

[52] Angelo Vistoli, *Notes on grothendieck topologies, fibered categories, and descent theory*, `http://arxiv.org/abs/math.AG/0412512`.

[53] Noson S. Yanofsky, *Algebraic theories in quantum field theories and quantum algebra: A proposal*, (1999), `http://www.sci.brooklyn.cuny.edu/~noson/pubs.html`.

[54] ———, *The syntax of coherence*, Cahiers Topologie Géom. Différentielle Catég. **41** (2000), no. 4, 255–304, `http://arxiv.org/abs/math.CT/9910006`. MR1805933 (2001h:18007)

[55] ———, *Coherence, homotopy and 2-theories*, $K$-Theory **23** (2001), no. 3, 203–235, `http://arxiv.org/abs/math.CT/0007033`. MR1857207 (2003d:18014)

# Index

## Editorial Information

To be published in the *Memoirs*, a paper must be correct, new, nontrivial, and significant. Further, it must be well written and of interest to a substantial number of mathematicians. Piecemeal results, such as an inconclusive step toward an unproved major theorem or a minor variation on a known result, are in general not acceptable for publication. Papers appearing in *Memoirs* are generally at least 80 and not more than 200 published pages in length. Papers less than 80 or more than 200 published pages require the approval of the Managing Editor of the Transactions/Memoirs Editorial Board.

As of March 31, 2006, the backlog for this journal was approximately 13 volumes. This estimate is the result of dividing the number of manuscripts for this journal in the Providence office that have not yet gone to the printer on the above date by the average number of monographs per volume over the previous twelve months, reduced by the number of volumes published in four months (the time necessary for preparing a volume for the printer). (There are 6 volumes per year, each containing at least 4 numbers.)

A Consent to Publish and Copyright Agreement is required before a paper will be published in the *Memoirs*. After a paper is accepted for publication, the Providence office will send a Consent to Publish and Copyright Agreement to all authors of the paper. By submitting a paper to the *Memoirs*, authors certify that the results have not been submitted to nor are they under consideration for publication by another journal, conference proceedings, or similar publication.

## Information for Authors

*Memoirs* are printed from camera copy fully prepared by the author. This means that the finished book will look exactly like the copy submitted.

The paper must contain a *descriptive title* and an *abstract* that summarizes the article in language suitable for workers in the general field (algebra, analysis, etc.). The *descriptive title* should be short, but informative; useless or vague phrases such as "some remarks about" or "concerning" should be avoided. The *abstract* should be at least one complete sentence, and at most 300 words. Included with the footnotes to the paper should be the 2000 *Mathematics Subject Classification* representing the primary and secondary subjects of the article. The classifications are accessible from `www.ams.org/msc/`. The list of classifications is also available in print starting with the 1999 annual index of *Mathematical Reviews*. The Mathematics Subject Classification footnote may be followed by a list of *key words and phrases* describing the subject matter of the article and taken from it. Journal abbreviations used in bibliographies are listed in the latest *Mathematical Reviews* annual index. The series abbreviations are also accessible from `www.ams.org/publications/`. To help in preparing and verifying references, the AMS offers MR Lookup, a Reference Tool for Linking, at `www.ams.org/mrlookup/`. When the manuscript is submitted, authors should supply the editor with electronic addresses if available. These will be printed after the postal address at the end of the article.

**Electronically prepared manuscripts.** The AMS encourages electronically prepared manuscripts, with a strong preference for $\mathcal{A}\mathcal{M}\mathcal{S}$-LaTeX. To this end, the Society has prepared $\mathcal{A}\mathcal{M}\mathcal{S}$-LaTeX author packages for each AMS publication. Author packages include instructions for preparing electronic manuscripts, the *AMS Author Handbook*, samples, and a style file that generates the particular design specifications of that publication series. Though $\mathcal{A}\mathcal{M}\mathcal{S}$-LaTeX is the highly preferred format of TeX, author packages are also available in $\mathcal{A}\mathcal{M}\mathcal{S}$-TeX.

Authors may retrieve an author package from e-MATH starting from `www.ams.org/tex/` or via FTP to `ftp.ams.org` (login as `anonymous`, enter username as password, and type `cd pub/author-info`). The *AMS Author Handbook* and the *Instruction Manual* are available in PDF format following the author packages link from `www.ams.org/tex/`. The author package can also be obtained free of charge by sending email to tech-support@ams.org (Internet) or from the Publication Division, American Mathematical Society, 201 Charles St., Providence, RI 02904-2294, USA. When requesting an author package, please specify $\mathcal{A}\mathcal{M}\mathcal{S}$-LaTeX or $\mathcal{A}\mathcal{M}\mathcal{S}$-TeX and the publication in which your paper will appear. Please be sure to include your complete email address.

**Sending electronic files.** After acceptance, the source file(s) should be sent to the Providence office (this includes any TeX source file, any graphics files, and the DVI or PostScript file).

Before sending the source file, be sure you have proofread your paper carefully. The files you send must be the EXACT files used to generate the proof copy that was accepted for publication. For all publications, authors are required to send a printed copy of their paper, which exactly matches the copy approved for publication, along with any graphics that will appear in the paper.

TeX files may be submitted by email, FTP, or on diskette. The DVI file(s) and PostScript files should be submitted only by FTP or on diskette unless they are encoded properly to submit through email. (DVI files are binary and PostScript files tend to be very large.)

Electronically prepared manuscripts can be sent via email to pub-submit@ams.org (Internet). The subject line of the message should include the publication code to identify it as a Memoir. TeX source files, DVI files, and PostScript files can be transferred over the Internet by FTP to the Internet node e-math.ams.org (130.44.1.100).

**Electronic graphics.** Comprehensive instructions on preparing graphics are available at www.ams.org/jourhtml/graphics.html. A few of the major requirements are given here.

Submit files for graphics as EPS (Encapsulated PostScript) files. This includes graphics originated via a graphics application as well as scanned photographs or other computer-generated images. If this is not possible, TIFF files are acceptable as long as they can be opened in Adobe Photoshop or Illustrator. No matter what method was used to produce the graphic, it is necessary to provide a paper copy to the AMS.

Authors using graphics packages for the creation of electronic art should also avoid the use of any lines thinner than 0.5 points in width. Many graphics packages allow the user to specify a "hairline" for a very thin line. Hairlines often look acceptable when proofed on a typical laser printer. However, when produced on a high-resolution laser imagesetter, hairlines become nearly invisible and will be lost entirely in the final printing process.

Screens should be set to values between 15% and 85%. Screens which fall outside of this range are too light or too dark to print correctly. Variations of screens within a graphic should be no less than 10%.

**Inquiries.** Any inquiries concerning a paper that has been accepted for publication should be sent directly to the Electronic Prepress Department, American Mathematical Society, 201 Charles St., Providence, RI 02904, USA.

# Editors

This journal is designed particularly for long research papers, normally at least 80 pages in length, and groups of cognate papers in pure and applied mathematics. Papers intended for publication in the *Memoirs* should be addressed to one of the following editors. In principle the Memoirs welcomes electronic submissions, and some of the editors, those whose names appear below with an asterisk (*), have indicated that they prefer them. However, editors reserve the right to request hard copies after papers have been submitted electronically. Authors are advised to make preliminary email inquiries to editors about whether they are likely to be able to handle submissions in a particular electronic form.


***Algebra** to ALEXANDER KLESHCHEV, Department of Mathematics, University of Oregon, Eugene, OR 97403-1222; email: ams@noether.uoregon.edu

***Algebra and its application** to MINA TEICHER, Emmy Noether Research Institute for Mathematics, Bar-Ilan University, Ramat-Gan 52900, Israel; email: teicher@macs.biu.ac.il

**Algebraic geometry** to DAN ABRAMOVICH, Department of Mathematics, Brown University, Box 1917, Providence, RI 02912; email: amsedit@math.brown.edu

**Algebraic number theory** to V. KUMAR MURTY, Department of Mathematics, University of Toronto, 100 St. George Street, Toronto, ON M5S 1A1, Canada; email: murty@math.toronto.edu

***Algebraic topology** to ALEJANDRO ADEM, Department of Mathematics, University of British Columbia, Room 121, 1984 Mathematics Road, Vancouver, British Columbia, Canada V6T 1Z2; email: transactions@math.ubc.ca

***Combinatorics** to JOHN R. STEMBRIDGE, Department of Mathematics, University of Michigan, Ann Arbor, Michigan 48109-1109; email: JRS@umich.edu

**Complex analysis and harmonic analysis** to ALEXANDER NAGEL, Department of Mathematics, University of Wisconsin, 480 Lincoln Drive, Madison, WI 53706-1313; email: nagel@math.wisc.edu

***Differential geometry and global analysis** to LISA C. JEFFREY, Department of Mathematics, University of Toronto, 100 St. George St., Toronto, ON Canada M5S 3G3; email: jeffrey@math.toronto.edu

**Dynamical systems and ergodic theory** to AMIE WILKINSON, Department of Mathematics, Northwestern University, 2033 Sheridan Road, Evanston, IL 60208-2730; email: transactions@math.northwestern.edu

***Functional analysis and operator algebras** to MARIUS DADARLAT, Department of Mathematics, Purdue University, 150 N. University St., West Lafayette, IN 47907-2067; email: mdd@math.purdue.edu

***Geometric analysis** to TOBIAS COLDING, Courant Institute, New York University, 251 Mercer St., New York, NY 10012; email: traneditor@cims.nyu.edu

***Geometric topology** to MLADEN BESTVINA, Department of Mathematics, University of Utah, 155 South 1400 East, JWB 233, Salt Lake City, Utah 84112-0090; email: bestvina@math.utah.edu

**Harmonic analysis, representation theory, and Lie theory** to ROBERT J. STANTON, Department of Mathematics, The Ohio State University, 231 West 18th Avenue, Columbus, OH 43210-1174; email: stanton@math.ohio-state.edu

***Logic** to STEFFEN LEMPP, Department of Mathematics, University of Wisconsin, 480 Lincoln Drive, Madison, Wisconsin 53706-1388; email: lempp@math.wisc.edu

***Ordinary differential equations, partial differential equations, and applied mathematics** to PETER W. BATES, Department of Mathematics, Michigan State University, East Lansing, MI 48824-1027; email: bates@math.msu.edu

**Partial differential equations** to GUSTAVO PONCE, Department of Mathematics, South Hall, Room 6607, University of California, Santa Barbara, CA 93106; email: ponce@math.ucsb.edu

***Probability and statistics** to KRZYSZTOF BURDZY, Department of Mathematics, University of Washington, Box 354350, Seattle, Washington 98195-4350; email: burdzy@math.washington.edu

***Real analysis and partial differential equations** to DANIEL TATARU, Department of Mathematics, University of California, Berkeley, Berkeley, CA 94720; email: tataru@math.berkeley.edu

**All other communications to the editors** should be addressed to the Managing Editor, ROBERT GURALNICK, Department of Mathematics, University of Southern California, Los Angeles, CA 90089-1113; email: transams@math.usc.edu.